Cambridge Elements ≡

Elements in Organizational Response to Climate Change
edited by
Aseem Prakash
University of Washington
Jennifer Hadden
University of Maryland
David Konisky
Indiana University
Matthew Potoski
UC Santa Barbara

CLIMATE ACTIVISM, DIGITAL TECHNOLOGIES, AND ORGANIZATIONAL CHANGE

Mette Eilstrup-Sangiovanni
University of Cambridge

Nina Hall
*Johns Hopkins University
School of Advanced International Studies (SAIS)*

CAMBRIDGE
UNIVERSITY PRESS

CAMBRIDGE
UNIVERSITY PRESS

Shaftesbury Road, Cambridge CB2 8EA, United Kingdom

One Liberty Plaza, 20th Floor, New York, NY 10006, USA

477 Williamstown Road, Port Melbourne, VIC 3207, Australia

314–321, 3rd Floor, Plot 3, Splendor Forum, Jasola District Centre, New Delhi – 110025, India

103 Penang Road, #05–06/07, Visioncrest Commercial, Singapore 238467

Cambridge University Press is part of Cambridge University Press & Assessment, a department of the University of Cambridge.

We share the University's mission to contribute to society through the pursuit of education, learning and research at the highest international levels of excellence.

www.cambridge.org
Information on this title: www.cambridge.org/9781009483506

DOI: 10.1017/9781009483544

When citing this work, please include a reference to the DOI 10.1017/9781009483544

First published 2024

A catalogue record for this publication is available from the British Library

ISBN 978-1-00948-350-6 Hardback
ISBN 978-1-00948-353-7 Paperback
ISSN 2753-9342 (online)
ISSN 2753-9334 (print)

Cambridge University Press & Assessment has no responsibility for the persistence or accuracy of URLs for external or third-party internet websites referred to in this publication and does not guarantee that any content on such websites is, or will remain, accurate or appropriate.

Climate Activism, Digital Technologies, and Organizational Change

Elements in Organizational Response to Climate Change

DOI: 10.1017/9781009483544
First published online: December 2024

Mette Eilstrup-Sangiovanni
University of Cambridge

Nina Hall
Johns Hopkins University
School of Advanced International Studies (SAIS)

Author for correspondence: Mette Eilstrup-Sangiovanni, mer29@cam.ac.uk

Abstract: Non-governmental and civil society organizations have long been recognized as crucial players in climate politics. Today, thanks to the internet, social media, satellite, and other technologies, climate activists are pioneering new organizational forms and strategies. Organizations like Fridays For Future, 350.org, and GetUp! have used social media and other digital platforms to mobilize millions of people. Many NGOs use digital tools to collect and analyze 'big data' on environmental factors, and to investigate and prosecute environmental crimes. Although the rise of digitally based advocacy organizations is well documented, we know less about how digital technologies are used in different aspects of climate activism, and with what effects. On this basis, we ask: how do NGOs use digital technology to campaign for climate action? What are the benefits and downsides of using technology to push for political change? To what extent does technology influence the goals activists strive for and their strategies?

Keywords: climate activism, digital technology, NGOs, climate change, organizational change

ISBNs: 9781009483506 (HB), 9781009483537 (PB), 9781009483544 (OC)
ISSNs: 2753-9342 (online), 2753-9334 (print)

Contents

Foreword   1

Introduction   2

1 Mobilizing, Organizing, and Campaigning   13

2 Monitoring and Enforcement   29

3 Lobbying   51

4 Forming, Fundraising, and Networking   58

5 The Dark Side of Digital Technologies   65

Conclusion   73

References   80

Foreword

This Element grew out of several years of conversations about climate change, advocacy, and the role of digital technologies. Between 2017 and 2022 we were both writing books about different forms of advocacy: Mette co-authored *Vigilantes beyond Borders* (with J. C. Sharman), which explored the role of non-governmental organizations (NGOs) in direct enforcement, Nina wrote *Transnational Advocacy in the Digital Era*, which examined how advocacy organizations use email, social media, and data analytics to mobilize thousands of people and put pressure on governments. While our books focused on different kinds of advocacy organizations and examined different strategics and tactics, we realized the potential to collaborate on understanding how new technologies interact with broader trends in climate advocacy. We wrote this Element with the aim of connecting disparate literatures, identifying new trends in use of technology by climate organizations and considering what questions these raise for scholars and practitioners. We thereby hope to chart a broad research agenda for scholars working at the intersection of digital politics and climate advocacy, social movements, and NGOs.

Given the fast pace of technological innovation (ChatGPT was a new tool when we started writing this Element in 2023), we do not try to cover every single relevant technology or novel use. Rather, as scholars with expertise in social and political *organizations*, we are interested in how NGOs, civil society organizations (CSOs), and other organized advocacy groups are adapting to new technologies, and what this means for their organizational structures, strategies, and goals. This organizational-level perspective is often missing from the field of digitalization and advocacy, which tends to consider techno-logical innovations independently of the organizations that use them. Our aim is thus to start a conversation and outline a research agenda focused on digital technologies and organizational change in climate advocacy.

Introduction

Non-governmental organizations and civil society organizations are widely recognized as important players in global climate politics. However, new technologies are changing how these organizations develop and operate. Empowered by social media, big data, artificial intelligence (AI), satellite, and more, activists across the world are experimenting with new forms of organization and new strategies to promote climate action. For example, digital advocacy organizations like Fridays For Future, 350.org, Campact, and GetUp! have used online platforms to mobilize millions of people in support of climate action (Hall, 2022). They have done so with fewer resources than traditional environmental NGOs such as Greenpeace and Friends of the Earth, who have rarely managed to rally support on this scale. Many climate groups today use digital platforms to fundraise, form alliances, educate global audiences, and connect with activists in remote locations. Others rely on digital technologies to collect, process, and analyse vast amounts of data, which is used to lobby policymakers, monitor companies, or investigate and prosecute environmental crimes committed by governments or corporate actors (Eilstrup-Sangiovanni and Sharman, 2022). At the same time, digital technologies also present new challenges for climate activism, including risks of 'slacktivism' and 'vanity metrics', growing state surveillance, rapid spread of misinformation, and new patterns of political exclusion based on uneven access to technological infrastructures.

Clarifying the uses, and pitfalls, of technology reliance for climate activists is crucial given the rapid evolution of digital technologies. Climate change is a defining challenge of our age, with time fast running out to avert catastrophic consequences (de Moor, 2021). Many NGOs and CSOs are stepping up their work in this domain, employing a wide range of strategies from research to raising public awareness, staging large-scale protests, lobbying policymakers, implementing and enforcing public policies, or working directly with vulnerable communities outside of policy venues. New digital (and also some non-digital) technologies play a growing role in all these activities. So far, the dominant focus in literature on advocacy and digital technology has been on how new information and communication technology (ICT) and social media platforms lower barriers to information exchange and facilitate wider mobilization and participant-led advocacy. Yet, as we show in this Element, the impact of technology on climate activism extends far beyond mobilization and member-driven campaigning to organizational formation and fundraising, data collection and research, lobbying, monitoring, and enforcement. For example, remote sensing and Geographic Information System (GIS) mapping are habitually used by activists to collect and analyze data on climate change indicators, such as deforestation, changes in land use, greenhouse gas (GHG) emissions, and glacier melting. Satellites, as well as drones and other unmanned aerial vehicles, are also used to monitor environmental problems in remote and inaccessible areas, like forests and oceans, where the effects of climate change can be difficult to document, especially for non-state actors. Meanwhile, many NGOs use advanced data analytics to process large and complex data, such as satellite imagery and other forms of geo-spatial information, which can be used to track emissions and air pollution, or to detect specific environmental crimes, helping activists to decide where to focus their advocacy or enforcement efforts (Eilstrup-Sangiovanni and Sharman, 2022).

Against this backdrop of widening technology use(s), this Element asks: how do advocacy organizations use new digital technologies to campaign for climate action? Specifically, we seek to understand how technological developments influence the formation and structure of climate organizations, the goals they strive for, and their choice of strategies and tactics.[1] Although the rise of digitally based advocacy groups that predominantly mobilize and engage their members online has been well documented (Hall et al., 2020), we know less about how digital technologies are used in other aspects of climate advocacy, and how different technologies may influence organizational structures and 'repertoires of action' (Tilly, 1977). To address these questions, we examine the role of digital technologies across several aspects of climate activism – from organizational formation and fundraising to alliance-building, networking, mobilizing and campaigning, elite lobbying, monitoring, and enforcement. Rather than focusing on a single phase of the 'advocacy cycle' (or a single type of organization), we thus examine how new technologies are impacting the structures and practices of climate activist organizations across a range of functions and organizational types.

In this Element, we focus predominantly on existing technologies, rather than speculating about future innovations. Our approach is neither optimistic nor pessimistic on technology; we examine the benefits as well as the risks of technology use for a range of climate advocacy activities and organizations. On the one hand, the growing speed and declining cost of online communication mean that activist groups can cheaply and easily connect with their members and other activists worldwide, facilitating cross-border networking, fundraising, mobilization, coordinated action, and mutual learning. Datafication and big data analytics also afford unprecedented access to data with which activists can hold states and corporations to account. On the other hand, potential downsides of digital advocacy are coming into clearer view. Both authoritarian and democratic regimes increasingly use technology to track, surveil, and repress climate

---

[1] Berry (1977) distinguished advocacy *strategies* – general, long-range approaches to influencing public policy – from advocacy *tactics* – specific actions taken to execute a particular strategy. Examples of tactics could be boycotts, research, media campaigns, lobbying, or expert testimony/litigation.

activists. Growing reliance on digital platforms to mobilize public support can lead to 'clicktivism' (that is, predominantly online engagement with political causes) or 'slacktivism' (that is, fleeting and superficial engagement with a cause) and to new patterns of political exclusion. Some of these pitfalls have been highlighted in existing literature on social movements and political communications (Gladwell, 2010). Yet, other downsides have not featured prominently in existing scholarship. For example, the increasing availability and declining cost of new digital technologies have led to a surge in organizational formation and strategic innovation in a relatively short space of time. Although this may entail benefits of greater scale and span of climate activism, it also means that new strategies and practices may proliferate without having been properly tested, leading to missteps and setbacks for campaigns. Another risk is that opponents harness the same technologies to campaign *against* climate action. We have seen, for example, the rise of online disinformation to undermine climate science and climate advocacy; and countermovements have used social media to organize and attack prominent climate activists. Thus, rather than assuming that digitalization supports climate activism, we seek to better understand how digital technologies facilitate new organizational forms and practices, and what specific opportunities and risks technology presents for climate activists.

This Element is divided into five sections that each address the role of technology in a different aspect of climate activism. In each section, we focus on opportunities and challenges afforded by new technology use and offer practical, illustrative examples from around the world. This broad-stroke, illustrative approach is not intended to offer a definitive overview or appraisal of technology use by climate activists or its effects. Nor are we able, in this short Element, to provide a thorough introduction to the growing literature that deals with digitally based advocacy and activism. Rather than provide a comprehensive introduction to climate advocacy in digital spaces, our goal is to draw attention to emerging trends, and identify critical questions for scholars. By examining the potential of new technologies to transform climate advocacy and by highlighting risks and challenges arising from technological innovation, we outline a broad research agenda for international relations and social movement scholars. We hope this Element will serve as a springboard for more systematic and in-depth research and be a useful guide for activists thinking about how to harness new technologies for climate advocacy.

More of the Same, Just Faster or a Transformative Impact?

Since the 1990s, scholarship on transnational advocacy has highlighted the benefits of modern ICT (e.g., Keck and Sikkink, 1998). Email, internet, and social media serve as virtual megaphones, enabling activists to broadcast their message to a global audience and fostering a sense of interconnectedness among disparate communities. In this literature, digital technologies are often credited with enhancing organizational capacities by rationalizing or speeding up familiar processes. Just like commercial organizations, CSOs and NGOs can exploit digital technologies to reduce costs, enhance efficiency of internal working processes, and extend the reach of their communications. For example, online petitions make the process of gathering signatures for a petition more efficient than the analogue version of this tactic but do not fundamentally change the tactic (Hestres and Hopke, 2016). Recruiting new supporters via email or social media is easier and cheaper than are door-to-door appeals or phone campaigns but, again, does not fundamentally change the practice.

However, technology can also have a more *transformative* impact on advocacy organizations. Digital technologies have the potential to fundamentally change the collective action process, by allowing individuals to eschew traditional brick-and-mortar organizations and use digital platforms to self-organize (Hestres and Hopke, 2016). Digital technologies can also enable new forms of political activism such as 'data activism' (Milan and Velden, 2016) or 'crowdfunding', and boost innovation and learning across organizations by increasing organizational capacity to receive and act on new information from outsiders (Hall et al., 2020). Finally, the rapid scaling of campaign activities and membership enabled by digitalization can have a transformative effect on organizational ecologies. For example, in the past few decades, new climate organizations have emerged whose scope and influence would be hard to imagine without digital technologies. This has threatened the role of traditional brick-and-mortar advocacy organizations (Hestres and Hopke, 2016) and forced them to innovate, for example, by increasing their online presence and seeking to engage their membership more directly (Schmitz et al., 2020).

Throughout this Element, we provide examples both of technology as a booster of long-standing practices and strategies, and of its transformative role as an instigator of new practices and meanings in the struggle to halt destructive climate change. Our main goal, however, is to identify cases and circumstances in which technology appears to be driving fundamental organizational change. In sum, our interest is less in how technologies enable groups to do familiar things faster or cheaper (although this is also important) and more in how technology may prompt groups to do things *differently*.

Aims and Scope

We are interested in how new technologies drive organizational change and innovation. By organizational change, we mean changes in organizational structures, missions, cultures, strategies, and tactics. As such our focus is not limited to a specific technology. Rather, we throw the net wide, looking at technologies from ICT and social media to big data analytics, AI, satellite, drones, and infrared cameras. We realize that different technologies may afford different opportunities and constraints for climate activists, and impact different groups differently. As such, we do not develop – let alone test – hypotheses about how specific technologies impact climate activism. Rather, our aim is to highlight these differences and identify questions for future research.

Just as we do not focus on a specific technology, we also do not limit our focus to a particular form of climate activism. The strategies adopted by climate activists are manifold and diverse (Hadden and Jasny, 2017; Eilstrup-Sangiovanni, 2019). Organizations that share similar goals often adopt widely different methods to achieve the change they want to see. Some organizations (like Worldwide Fund for Nature (WWF)) collaborate directly with industries to develop eco-friendly certification schemes, while others focus on dragging dirty corporations before courts (e.g., ClientEarth). Greenpeace stages public protests outside major energy companies, while the Union of Concerned Scientists produces detailed scientific reports on GHG emissions (Hadden and Jasny, 2017). Some groups seek to persuade policymakers through lobbying or evidence-based policy advice, while others (such as 350.org) mobilize citizens to 'rise up' and demand change from people in positions of power. These different approaches reflect different beliefs about what kinds of interventions are most likely to spark change or, in the language of practitioners, different 'theories of change'.

At its simplest, a theory of change represents an overarching strategy, or organizational logic, for how an organization seeks to realize the goal(s) it campaigns for.[2] Theories of change describe how a specific activity (say, a demonstration or court case) may lead to a specific goal (say, changing a given country's climate policy) by stipulating the assumptions and context the organization is operating in, including what actors are instrumental to producing change. For example, organizations that engage in climate litigation argue that the courts can be effectively used to hold states (or private companies) to account for their climate actions (or lack thereof). Meanwhile, organizations like Last Generation see direct action as the most effective strategy: only by physically stopping fossil-fuel-powered traffic on roads and at airports can we shift public opinion and change government policy. Some activist organizations subscribe to just one theory of change (e.g., Last Generation mostly engages in direct action) while others work with multiple theories (Greenpeace engages in direct action, litigation, street marches, lobbying, and gathering of scientific evidence). Importantly, different theories of change may encourage different uses of technology. For example,

---

[2] Theory of Change Institute, 'What Is a Theory of Change', www.theoryofchange.org/what-is-theory-of-change/.

groups that engage in monitoring and enforcement often employ satellite, remote sensing, and big/open data, whereas groups that emphasize grassroots mobilization rely more strongly on social media. The social media strategies of organizations that focus on public mobilization in turn differ from those of organizations that emphasize elite persuasion (Hestres, 2015).

The sections of this Element correspond broadly to different theories of change. Table 1 summarizes the main theories covered. In each section, we offer illustrative examples of technologies and organizations that exemplify these distinct theories of change. The list is not exhaustive – there are theories of change we do not cover here that are relevant to climate activism – nor are the featured theories and associated strategies mutually exclusive; as noted, many advocacy organizations successfully combine different theories of change.

What groups and organizations are we looking at? Our focus is on non-state climate activist organizations. We define activism as 'actions aimed at fostering, obstructing or guiding political and environmental change' (Gutiérrez, 2018). By non-state actors we mean NGOs, CSOs, and independent scientific institutions operating on a non-profit basis. Given the complexity of climate change, many NGOs and CSOs working to promote biodiversity, sustainable development, and human rights incorporate climate goals into their broader portfolio of work. To paraphrase Brett Solomon,[3] 'you can't run a 21st century NGO without some focus on climate'. Our focus, however, is on organizations whose *primary* aim is to encourage climate action, whether through reducing GHG emissions, protecting vulnerable forest and marine environments to increase carbon storage, or mitigating the effects of climate change on vulnerable populations. Within this population, we adopt a 'transcalar' approach to examine climate activist organizations at the local, national, regional, transnational, and international levels (Pallas and Bloodgood, 2022). We look at groups that adopt a range of ideological positions, goals, strategies, and tactics, whether 'inside' or 'outside' the established political system. This broad focus allows us to identify different innovative uses of technology at different scales, and in different political and geographical contexts.

Importantly, we include both 'digitally native' organizations and those that existed before the digital/internet era. The contemporary world is not clearly demarcated between digital and non-digital spheres. Some of the most attention-grabbing climate activism in recent times has relied on traditional tactics like street marches, blockades, and direct actions such as activists gluing themselves to streets. Meanwhile, digital tools are often integrated into on-the-ground protests whose impact is amplified by hashtag campaigns, online petitions, and virtual events. Rather than online mobilization replacing traditional

---

[3] Interview with Brett Solomon, 28 November 2023.

Table 1 Theories of change.

Theory of change	Rationale	Tactics	Organizational examples	Technology	Section
Mobilizing the public	Mobilizing public opinion to pressure elected officials or other powerful actors to enact change	Petitions; protest marches; media campaigns; leafleting; 'naming and shaming'; public forums	Fridays For Future; Extinction Rebellion; Last Generation; Sunrise Movement; Greenpeace 350.org Sierra Club	Social media; email; websites; online petitions; AI	Section 1
Bottom-up, grassroots organizing	Organizing and empowering individuals to advocate and enact change on their own behalf	Community organizing; working with groups of 'concerned citizens' to develop capacity	350.org; Waorani; Digital Democracy; Indigenous Mapping Collective; Tierras Indígenas; Cadasta Foundation; Project Canopy	Social media; email; radio; GPS: GIS; satellite	Sections 1 and 2
Monitoring compliance and general behaviour	Monitoring compliance with laws, policies, and pledges	Emissions tracking; undercover investigations	Climate Action Tracker; ClimateTRACE; SkyTruth; Joint Impact Model	Satellite; GIS; open-source data; AI; mobile phones; citizen-science platforms	Section 2

Enforcing domestic and international law	Ensuring that existing national laws and international climate agreements are respected	Litigation; legal advice; court hearings; submissions of evidence; direct action	Global Witness; Greenpeace; ClientEarth; SkyTruth	Satellite; GIS; open-source data; geo-spatial mapping	Section 2
Direct action	Taking action to stop fossil-fuel infrastructure or GHG emissions directly at source	Blockades; boycotts; civil disobedience; divestment campaigns	Last Generation; Greenpeace; Sea Shepherd	Email; social media; cellphones	Section 2
Elite decision-maker persuasion	Directly influencing government/key decision-makers, or corporate actors through lobbying and policy advice	Lobbying; letter writing; petitioning, in-person meetings; presenting scientific evidence; policy analysis and research	Influence Map; Climate Cabinet Education; Climate Policy Radar	Email; social media; open-source data	Section 3
Producing and disseminating scientific evidence	Informing policy through scientific evidence	Research and policy-maker education	Zooniverse; Climateprediction.net	Crowdsourcing; citizen-science platforms	Section 2

Note: This table does not cover all theories of change. Others include influencing individual behaviours to reduce emissions; shareholder activism; setting voluntary standards for and with the private sector; capacity-building; and campaigning for political candidates.

offline protests, online and off-line tactics thus increasingly blend in ways that call for further analysis.

t is important to stress that the organizations we examine have not been selected to be representative of the general population of NGOs or CSOs, nor do they constitute traditional case studies selected to test specific hypotheses. We are interested in examples that highlight and illustrate the transformative potential of current technologies rather than in showcasing conventional or typical uses.

Not all the technologies we focus on are digital and not all are strictly speaking new (Karpf, 2020). Machine learning is not new (depending on how we define 'new') and neither is the internet, which has been with us since 1969. We are interested in how new *uses* of technology are changing climate advocacy. For example, how social media enable climate groups to mobilize on a wider scale, or how satellite imagery and remote sensing help change the way many people perceive climate change – not merely through producing scientific data that demonstrate that climate change is 'real', but also by turning what many might perceive as an abstract planetary issue into something more concrete and tangible; for example, by clearly linking local weather events to global atmospheric changes.

Some climate activists are sceptical about the positive contribution of technologies to climate change and advocate for a return to simpler practices in agriculture, transportation, and consumption.[4] Given the focus of this Element – how technology is transforming climate advocacy – we do not explore the perspectives of climate activists that are opting out of using technology altogether. However, as we discuss further in Section 5 on 'the dark side of digital technologies', there is an important strand of tech-sceptic climate activism which questions the environmental costs of technology and attendant economies of 'surveillance capitalism' (Crawford, 2021).

Drawing Together Existing Literature

Despite a growing literature on climate activism in the fields of international relations, social movement studies, sociology, environmental studies, geography, and digital and political communications studies, there is a dearth of scholarship examining how technology is transforming specific aspects of climate activism. There is a well-established literature on the use of digital technologies by NGOs for general information-sharing, mobilization, fundraising (Howson, 2021), recruitment, and campaigning. Notions like 'Advocacy 2.0', the changing face of advocacy (Chalmers and Shotton, 2016), and web-supported activism (Neumayer and Svensson, 2016) have been put forward, alongside concepts like 'slacktivism' (Gladwell, 2010). There is also an

---

[4] We thank Jean-Frédéric Morin for highlighting this point.

extensive literature on climate movements (e.g., Dietz and Garrelts, 2014; Fisher, 2019; Fisher and Nasrin, 2021). However, few scholars have specifically examined the impact of digital technologies on climate advocacy (notable exceptions are Hestres, 2015; and Dauvergne, 2020).

At the same time, literature on social movements and advocacy has not examined, or compared, the role of technology across different aspects and functions of advocacy, or different types of advocacy organizations. To the extent they have focused on technology use, studies have tended to focus on discrete technologies, such as satellite (Aday and Livingston, 2016; Rothe and Shim, 2018) or social media (Hall et al., 2020); or on specific strategies such as digital analytics (Beraldo and Milan, 2019). Empirically, the research agenda has been dominated by cases from North America (Cheon and Urpelainnen, 2018; Fisher and Nasrin, 2021) and Western Europe (de Moor et al., 2020; see also Baran and Stoltenberg, 2023). To fill these gaps, this Element considers a wider range of climate-focused NGOs and CSOs based in both the Global North and South. We do not argue that climate activism is distinctive in its use of digital technologies. Many different activist organizations and movements use technology to campaign – from the human rights movement to anti-corruption organizations (Eilstrup-Sangiovanni and Sharman, 2021, 2022). However, we believe climate advocacy offers a particularly fertile domain for exploring the intersection of technological and organizational change because we have seen so many new climate organizations forming, and new strategies and tactics being used. The high pace of organizational innovation, alongside the high political and public salience of climate change, make this an important area to explore for scholars of transnational advocacy.

Who Uses Tech for What?

Studies of transnational advocacy organizations often invoke dichotomies such as large versus small organizations, old versus young, resource-rich versus resource-poor, and hierarchical versus decentralized movements (Dellmuth and Bloodgood, 2019; Eilstrup-Sangiovanni, 2019). A common assumption is that technology use and capacity for innovation differ across advocacy organizations, with smaller and newly founded organizations generally being more tech-savvy than older, more established groups. For example, previous studies have found that NGOs founded before the digital media age often struggle to embrace transformational change because it challenges their staff-led, and expertise-driven organizational structures (Hall et al., 2020). One activist we spoke to suggested that youth groups have been able to 'adapt more quickly to new social media'.[5] Other studies have

---

[5] Interview with Michael Poland, Campaign Director, Fossil Fuel Non-Proliferation Treaty Initiative, 10 February 2023.

found that older and larger legacy organizations often have more resources to invest in digital strategies and in building a strong presence online (Hall et al., 2020; Hong et al., 2020). Many legacy NGOs view digital tools as a relatively inexpensive way of expanding their support base and disseminating information to members. In contrast, younger, digitally native groups are more likely to use digital tools to communicate directly with supporters, seeking input through A/B testing or other forms of digital analytics (Hall et al., 2020).

A similar question regarding relative organizational advantages applies to the use of digital technologies for research and for monitoring and enforcement. While the growing availability and plummeting costs of remote-sensing technologies such as infrared cameras and drones present new opportunities for data-collection and environmental monitoring by NGOs operating on a tight budget, activism based around use of satellite imagery and big data analytics is often more resource-demanding and may therefore appeal more to NGOs with strong organizational and funding structures. This raises the question whether some organizations are better positioned to leverage new technologies than others, or whether different groups derive different benefits from technology use. There may also be important differences in how groups harness technology according to geographical and political factors. With the question 'who uses tech for what purposes', we engage with organizational ecology perspectives to understand patterns of specialization and 'niche-seeking' among advocacy groups (e.g., Eilstrup-Sangiovanni, 2019; Eilstrup-Sangiovanni & Sharman, 2022; Bush & Hadden, 2019).

Book Roadmap

This is an Element about how climate activists across the world are harnessing new technologies. This does not mean that the technologies and uses we point to are necessarily unique to the domain of climate activism. However, given the magnitude of action in this space, we expect technological innovation to be particularly visible there.

In Section 1 we examine the role of digital technologies in mobilization and campaigning. ICT and social media have provided climate activists with unprecedented tools for mobilization, communication, and campaigning. Social media platforms allow activists to broadcast their messages to wider audiences, test new campaign issues and framings, converse directly with their members, and broker large digitally distributed movements – often with global reach. Recently, many advocacy organizations have experimented with using AI to generate content and to enhance the efficacy of their communication. However, reliance on digital platforms to engage supporters also comes with dangers, including clicktivism, slacktivism, 'vanity metrics', and increased risks of state surveillance.

Section 2 considers how digital technologies beyond social media and ICT – for example, satellite, drones, geographic information systems, 'big data', and machine learning – enable new forms of environmental monitoring and enforcement. The focus here is on how digitally-enabled monitoring is facilitating a range of new 'outside' strategies; from direct action to corporate 'naming and shaming' campaigns, and litigation.

Section 3 examines how new technologies also support so-called 'inside' strategies. These stategies include elite lobbying directed at policymakers or large corporations, along with policy analysis and scientific research.

Section 4 explores how digitalization has changed the process of creating, maintaining, and funding climate activist organizations. We explore how the internet and social media platforms have made it easier both to establish and fundraise for climate organizations, and to coordinate action within and across groups.

Finally, section 5 turns to the 'dark side' of technology by considering how it can be used to surveil and repress activists, and how technology development and use itself contributes to climate change and environmental destruction through high energy use and reliance on scarce minerals. We also examine the rapid rise of online disinformation and problems of uneven access to digital infrastructures which produce new 'digital divides' both within and across countries.

The concluding section summarizes the bigger themes of the Element: Is technology changing climate activism, and making it more (or less) effective? Who uses technology for what? What are the pressing questions for scholars and practitioners trying to understand how digital technology is transforming advocacy and how best to harness its potential while avoiding its pitfalls?

Our analysis in each section draws on a combination of desk research and interviews with climate activists and technology consultants who support activist organizations. Interviews were predominantly used to provide a background understanding of how activists experience and engage with technology. As a result, some interviews may not be directly cited in the text.

1 Mobilizing, Organizing, and Campaigning

In the lead-up to the 2009 United Nations Framework Convention on Climate Change (UNFCCC) summit in Copenhagen, 350.org, a new climate advocacy organization, founded by university students at Middlebury college and veteran climate activist Bill McKibben, led a Global Day of Action. Thousands took to the streets around the world, demanding climate action. CNN described it as 'the most widespread day of political action in our planet's history'.[6] What was striking was that a student-led initiative managed to rally so many people across the globe. As

---

[6] 350.org, '10 years', https://350.org/10-years/ (last accessed 14 May 2021).

Bill McKibben explains, 350.org initially had 'seven college students, and no money, no organization, no lists or anything, just this thought that we would go out and try to do the work of building a global movement that hadn't been there before. Which, of course, was a ludicrous idea. There were seven students and seven continents, so each took one. The guy who took Antarctica also had to take the internet'.[7]

How did this tiny student organization achieve such massive mobilization? In a matter of months, coordinators at 350.org were able to leverage digital tools to mobilize climate activists and environmental organizations around the world to participate in their International Day of Climate Action.[8] They recruited members via email and encouraged them to form their own local chapters and organize local actions. Chapters formed in cities and towns across the world from New Zealand to Poland, South Africa to India. While local sections were largely left to their own devices, 350.org centrally coordinated Global Days of Action to ensure greater impact. Local sections were encouraged to organize actions on the same day, with the same ask, and register their events on a central online events page thus making the impact global. While the 350.org central team coordinated the timing of actions (usually demonstrations) and the key messages, they allowed anyone, anywhere to set up their own local section, organize an action and register their planned demonstration on a global map.[9]

Mass mobilizations by 350.org are not a lone example. Other grassroots organizations, such as Fridays For Future and Extinction Rebellion (XR), have also mobilised thousands, if not millions, of climate protesters globally. While these activist organizations have gained visibility and influence thanks to traditional offline tactics (e.g., marches, blockades and direct action) digital technology has enabled them to quickly scale up and become global movements. Like 350.org they have adopted a form of digitally distributed activism based on a centrally defined set of goals, and distributed power to members to set up their own chapters and organize activities, which has been critical to their success. Both Fridays For Future and Extinction Rebellion have global events maps, which enable them to combine local actions into a truly global movement. As one activist explained, they are 'not just in your capital city, London, Paris. It's in every city, village and every single Friday'.[10] The success of this model has been noticed by many larger professional NGOs, and some, such as Greenpeace, subsequently have tried to replicate this model of digitally distributed campaigning (Hall et al., 2020).

This section explores how technology such as social media, digital analytics and AI have played a pivotal role in enabling climate organizations to engage

---

[7] Bill McKibben, on file with authors.   [8] Interview with Jon Warnow, 22 November 2023.
[9] 350.org, 'Get Involved', https://350.org/get-involved/ (last accessed 22 March 2021).
[10] Interview with Nicolas Haeringer, 350.org, June 2019 (cited in Hall, 2022: 183).

in public-facing campaigns, and specifically to mobilize and organize members. A large literature focuses on how the internet has enabled social movements to increase their scale and geographic reach. As we illustrate here, this has important implications for which organizations – well-resourced, professionally staffed NGOs, or relatively resource-poor digital newcomers – can coordinate the largest, most effective climate actions. Digging one step deeper, this section illustrates how NGOs use social media and other digital technologies not only to *broadcast* their messages, but also to *listen* to their members, *test* new campaigns, and involve members in designing campaigns (Hall et al., 2020). These strategies can transform the relationship between the public/volunteers and staff of an advocacy organization, offering power to members to initiate their own campaigns.

Digital Mobilizing and Organizing

A crucial aspect of environmental activism is to raise public awareness and shape political agendas through 'information politics' (Keck and Sikkink, 1998). New communication technologies offer clear benefits to groups trying to reach large audiences and mobilize and organize them to push policymakers. Social media platforms, email, and mobile texts have made it vastly easier, cheaper, and quicker to run large-scale, public campaigns. Rather than relying on door-to-door canvassing, word of mouth, or distribution of pamphlets, activist organizations can reach thousands – if not millions – of people just by writing a Facebook post or sending a WhatsApp message or text. Social media also lowers the cost and expands the reach of public education activities (Margetts et al., 2015). Importantly, social media can enable climate groups to bypass traditional print and television media where attention is often captured by legacy NGOs, or by groups close to government (Thrall et al., 2014).

Students of social movements, political communications, and digital politics have explored the consequences of the internet for collective action (Bennett and Segerberg, 2013; Fung, 2022; Fung et al., 2013; Farrell, 2012). This literature speaks to broader trends in advocacy beyond the climate movement. Alongside this, there is a burgeoning literature on digital environmental activism. A recent literature review of 138 publications on the subject found that more than a third were published in the last two years (see Figure 1) (Baran and Stoltenberg, 2023). The authors found that recent studies tend to focus on the role of social media in enabling the growth of new social movements and on issue 'framing' rather than on the reach or impact of online communications. Importantly, many social media studies focus on the role of 'politically engaged individuals' (e.g., YouTube commentators or TikTok users) rather than climate advocacy organizations per se.

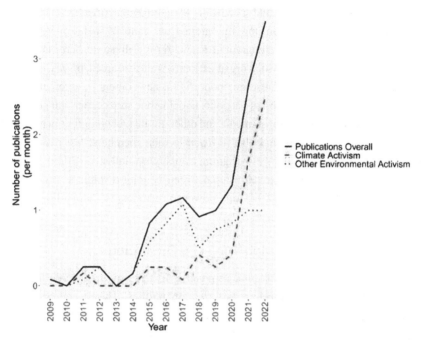

Figure 1 Scholarship on digital environmental/climate activism.
Source: (Baran and Stoltenberg 2023)

Much of this recent literature focuses on the role of digital technologies in 'supersizing' climate advocacy – that is, increasing the scale of advocacy efforts – rather than its transformative impact on how advocacy organizations operate. This is partly because many studies are interested in new forms of 'connective action', where people use digital platforms to build personalized rather than collective action frames and thereby increase mobilization (Bennett and Segerberg, 2013). In this mode, conventional political organizations, such as NGOs or CSOs, are no longer necessary for collective action, as political messages travel over social networking platforms, email lists, and online coordinating platforms (Bennett and Segerberg, 2013:742). There is little cost for the individual to engage in action, and minimal organizational coordination is needed, as the internet can connect disparate, autonomous activists, (Castells, 2012; Nyabola, 2018; Shirky, 2008:22). This trend towards 'organizing without organizations' implies that 'formal organizations are losing their grip on individuals and group ties are being replaced by large-scale, fluid social networks' (Bennett and Segerberg, 2012:748). Hence literature on digital 'connectivity' often does not explore how *formal* organizations can harness digital technologies to build global climate movements, or to change their modes of operation.

In contrast to scholarship highlighting the potential of 'organizing without organization', other scholarship finds that hierarchy, bureaucracy, and division of labour are essential for effective digital activism (Schradie, 2015; 2019; Han, 2014). Schradie (2019) compared grassroots and hierarchical advocacy organizations (albeit not climate orientated) in the United States and found that 'large hierarchical organizations can amplify their digital impact, whereas horizontal volunteer groups tend to be less effective at translating good-will into meaningful action'.[11] Her central point is that you need resources, expertise, and a clear division of labour to organize effectively online. Experienced activists and campaigners have reinforced the argument that centralized campaigns, led by professional advocacy organizations, and backed by professional staff who delegate action to members and volunteers, are needed to campaign effectively online (Bond and Exley, 2016; Mogus et al., 2011; Mogus and Liacas, 2016).

Four Digital Strategies for Climate Advocacy Organizations

A critical question is then: how do advocacy organizations harness digital technology to mobilise and organize the public for climate action? We find important variation in the digital strategies NGOs adopt, influenced by their different theories of change. Here we look not only at the role of social media in supersizing campaigns, but also how technology can transform the relationships between NGO staff and the broader public (Hall et al., 2020; Kingston and Stam, 2013:84). Hall et al. (2020) identify at least four potential ways that NGOs can harness digital technology in public campaigns (see Table 2). They can use it (1) to *broadcast* staff-generated messages; (2) to *test* new campaign issues and frames; (3) to *converse* with their members; or (4) to *facilitate* and broker large digitally distributed movements (as with the example of 350.org).

Rather than focusing on individual activists, or the social media platforms they use, this framework centres on the role of advocacy organizations and how

Table 2 Classifying NGO's digital advocacy strategies.[12]

	Staff-produced advocacy	Supporter-produced advocacy
Expand participation (breadth of engagement)	Broadcasting	Testing
Deepen commitment to cause (depth of engagement)	Conversing	Facilitating

---

[11] www.hup.harvard.edu/books/9780674972339.   [12] Adapted from Hall et al. (2020).

much power they are willing to devolve to members to determine campaign goals and strategies. This is important, as some scholars have suggested that organizations that *centralize* issue selection and *decentralize* implementation of campaigns are more likely to be successful (Wong, 2012). Importantly, each of the four strategies may use a range of technologies, including email, social media, AI, and peer-to-peer messaging. The question for scholars is not which platforms advocacy organizations are using, but rather *for what purpose*.

Broadcasting is the least transformative strategy. Here, NGOs simply use social media, email, or peer-to-peer texting to communicate what they would otherwise have done via letter, newspaper, radio, or television. This strategy is also the most studied, as scholars can relatively easily observe the online communications of advocacy organizations. There is variation in the platforms NGOs use to broadcast. In the Global North, NGOs and CSOs tend to rely more on Facebook, Twitter, and email, while in the Global South they gravitate towards WhatsApp/text/peer-to-peer messaging. In China, WeChat is the most popular application with 1.26 billion monthly active users in 2021 (Xu and Zhang, 2022). Meanwhile, groups in India, South Africa, and Kenya have pioneered the use of text and WhatsApp messaging for communicating with their members, and campaigning (Hall, 2022). They have thus 'leapfrogged' the stage of communicating via email (which most US and European organizations initially did).

In contrast to broadcasting, there is far less literature on how advocacy organizations engage in *testing* or 'analytic activism' (Karpf, 2016). Some advocacy organizations track which emails members open, and which have the highest action rate. Analytic activism was pioneered by digital advocacy organizations, such as MoveOn (in the US), GetUp! (Australia) and Avaaz (international) that work on multiple issues including climate change (Hall et al., 2020). Other climate-specific advocacy organizations have also begun to conduct digital analytics, as we illustrate later in this section. They analyse data on the performance of their social media posts to identify which campaigns and frames are most effective, and use this information to guide campaign choices. For example, senior campaign staff will meet regularly to discuss which petitions and campaigns are performing best and invest more staff energy and resources into these and drop campaigns which are performing poorly. They will also tweak the wording of mass emails based on which subject line is more effective and segment their membership depending on who engages most on which issues. The logic is to optimize engagement by members by identifying what colour, font, text, subject line, and issues are most likely to grab their attention (Karpf, 2016).

Climate advocacy organizations can also use technology to *converse* with their members. This is most easily done via social media, where organizations may solicit feedback, seek input on a post, and engage in discussion about issues to deepen engagement with members. Finally, organizations may give power to members to initiate their own campaigns via online petition platforms, and/or launch their own actions via an events page. An advocacy organization's role here is to *facilitate* a movement rather than dictate action to its members. Facilitating is potentially the most transformative of the four digital strategies as it involves handing over power to members to decide which campaigns to run and giving them the platform and power to do so. Greenpeace, for example, has set up online petition pages where any of their members can launch their own campaigning (Hall et al., 2020). The critical point is that digital technology enables new organizational practices based on listening, conversing, testing, and devolving power to a broader membership or public rather than concentrating it in the hands of professionalized staff. As such, new technologies can enable new forms of decision-making within climate advocacy organizations.

Fridays For Future

Fridays For Future (FFF) offers an excellent example of the different ways advocacy organizations use digital platforms to campaign. The movement claims 14 million members and has dozens of chapters around the world (they typically have a country-level organization – e.g., FFF Germany, FFF Nigeria – as well as sub-national chapters based in local cities or towns).[13] Fridays For Future have been extremely effective at mobilizing people and capturing media attention, and have adeptly used digital technology to coordinate large marches (Tattersall et al., 2022). Figures 2 and 3 show the number of countries and individuals involved in Fridays For Future marches in the first year of the movement's existence (Fisher and Nasrin, 2021:5). Remarkably, these began with Greta Thunberg's single-person demonstration in Sweden in September 2018 and culminated in the first global climate demonstration in March 2019 with 1 million participants. In September 2019, Fridays For Future organized one of the largest global demonstrations ever, with an estimated 7.3 million participants from 183 countries, mobilized over two consecutive Fridays. Although it is difficult to attribute specific policy or legislative changes to these large protests, representatives of the movement have gained direct access to government decision-makers, as well as corporations and international organizations, such as the World Economic Forum, and the United Nations (Fisher and Nasrin, 2021:5).

Importantly, Fridays For Future has a distributed model, which means that Greta Thunberg does not 'run' the movement or define its policies; rather each national chapter has significant autonomy to develop its own messages and

---

[13] Fridays For Future, https://fridaysforfuture.org/

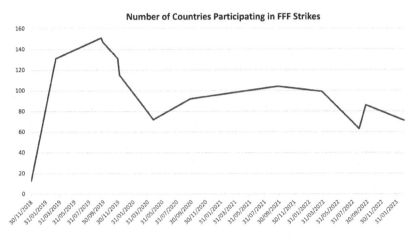

Figure 2 International growth in climate strikes by countries participating.

Source: https://fridaysforfuture.org/what-we-do/strike-statistics/list-of-countries/

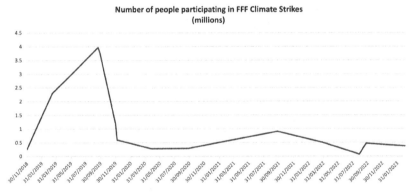

Figure 3 International growth in climate strikes by individual participants.

Source: https://fridaysforfuture.org/what-we-do/strike-statistics/list-of-countries/

actions (Nakabuye et al., 2020). Digital technology has enabled the movement to give members a platform to organize locally and yet be part of a global movement. The events map, found on Fridays For Future's website, is a crucial part of building this global climate movement (see Figure 4) as it unites and organizes disparate groups.

In addition to *facilitating* a global movement, we can see different ways in which Fridays For Future activists use digital platforms to maximise the reach of their online communications. For example, they also use social media to *broadcast* their work (Baran and Stoltenburg, 2023). As a US activist explained,

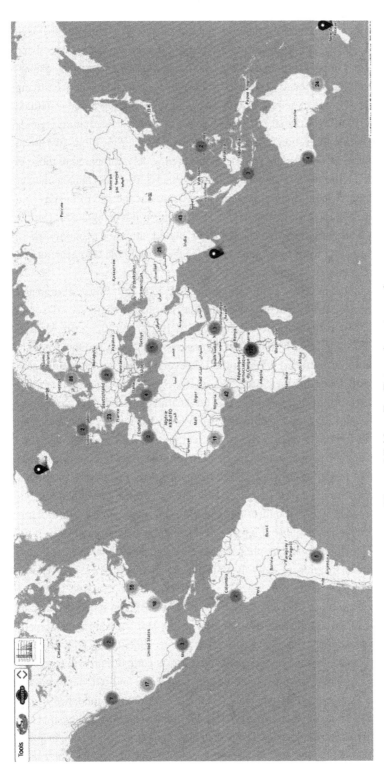

Figure 4 Fridays For Future (events map).

Source: https://map.fridaysforfuture.org/map?c=&d=&o=.

'Social media is our main – sometimes our only way – of getting the message out' (Sorce, 2023: 221). Many chapters have highlighted that algorithms shape their reach and impact on social media. As the same US-based activist elaborated: 'where [social media] sends that [message] . . . outside of maybe buying targeted ads . . . that is completely up to the algorithm. So, our reach is defined by the algorithm' (Sorce, 2023:221). Meanwhile, Fridays For Future Uganda found that their online content was less visible due to their location: 'We are in Africa . . . most of our content are [sic] not shared widely compared to those in Europe. That's the biggest challenge' (Sorce, 2023:220). The Uganda chapter also noted that some of their posts went unnoticed: 'At the end of the day you feel like you're not really . . . you feel like your voice is not being heard' (Sorce, 2023:221). This quote suggests that there is a 'digital' attention economy where content originating from some geographic locations (e.g., the US) gets noticed more than others (e.g., Uganda).

Although there have been multiple studies of Fridays For Future (Baran and Stoltenburg, 2023), we found only one that compared their use of social media, and algorithms across countries, namely Sorce (2023). In Austria, social media is handled by 'channel managers' (i.e., a set group of people responsible for a platform) while in the US, everyone in the leadership team can post to all social media (Sorce, 2023:218). Some activists explained that they try to use digital analytics to inform their campaigns and generate content specific to each social media platform. As an India-based activist stated: 'You just have to figure out what type of content suits which app and get your teammates and the people you know . . . the maximum people from the outside to interact so it will grow' (Sorce, 2023:221). Fridays For Future India also tracks the performance of their posts, using an 'in-app insights function', but activists noted it was difficult to use these digital analytics to inform campaigning decisions. As they explained: 'So I know the reach of my Insta[gram] page has gone down by 20 people but what does that actually mean? And how do I use those insights to actively improve my performance? That is still a mystery for me' (Sorce, 2023:222).

In contrast to other chapters, Fridays For Future Austria receives external support for its digital analytics from social media and software development agencies, who donate their time and skills (Sorce, 2023:222). This includes detailed analysis of social media productivity (posts per day); growth (percentage per week of new members) and engagement (total and individual posts). This data is discussed in weekly team meetings to 'future fine-tune their social media mobilization outreach' (Sorce 2023, 222). This is an excellent example of digital analytics in practice – where organizations look to optimize their messages for their audience and for specific social media platforms. There are, however, trade-offs for Fridays For Future Austria. On the one hand, they may be more effective compared to other national chapters that lack similar digital analytics expertise. On the other hand, they may be more dependent on private tech companies unless they find a way to develop this expertise in-house. More research is needed about how widespread analytic activism is among climate advocacy organizations and what enables and blocks organizations from using it.

AI, Chat GPT and Climate Campaigning

AI offers climate activists novel opportunities to scale up and speed up mobilization. Several activists we spoke to said they were still figuring out how AI, and particularly ChatGPT, may be useful. We also saw evidence of experimentation. We can use the four digital strategies outlined in Figure 4 to examine AI's potential effects for mobilizing, organizing and campaigning. First, advocacy organizations could use ChatGPT to *broadcast* messages more efficiently. For example, they could let ChatGPT draft emails to members and produce other content for campaigns, including images, videos, drafts of public submissions, and/or emails to politicians. AI could be particularly useful for organizations that wish to generate content quickly to catch a particular news cycle. Exemplifying this approach, *AI Impact Lab* advocates the use of AI in advocacy organizations and provides training to help organizations integrate AI into their campaigns to, for example, produce campaign material or donor reports.[14] The use of AI to produce content quickly and cheaply is what we most commonly came across in our research.

AI tools can also be used to *converse* with members and the wider public. Many of us are familiar with conversing with chatbots in other aspects of our lives, but to our knowledge, this function has not yet been used by climate activists. ChatGPT, for example, could help climate organizations collect and collate large amounts of member input and identify common concerns. However, it is important to note that large language models (LLMs) do not work well in all languages and dialects. An Indian climate activist explained to us how they used ChatGPT to summarize focus group and interview transcripts with marginalized communities in Maharashtra. They conducted these discussions in local dialects, but ChatGPT could not capture the differences in Indian dialects well.[15] Hence, at the time of our interview in late 2023, ChatGPT was of limited use in this context, although ChatGPT's capabilities are changing quickly.

---

[14] AI Impact Lab, https://aiimpactlab.com/about-1.

[15] Interview with Indian activist, 16 October 2023. Although another interview suggested AI is an excellent translation tool even for dialects.

While AI can help organizations to generate content faster and more cheaply, some activists we spoke to raised concerns about whether the material would be seen as inauthentic. Jon Warnow, former 350.org digital campaigner, explained that, although it is still early days in terms of generative AI producing more compelling content, 'the value of that written content is going to decline fairly precipitously in the next few years' as decision-makers will realize the content is AI produced. Another risk is that AI may replace staff in advocacy organizations. If AI can generate campaign material more effectively, and quickly, it will reduce staff costs. However, there are significant risks to reliance on AI given the biases present in AI algorithms, and the ongoing need for expertise to ensure accurate campaign claims. In addition, AI could pose privacy concerns related to how information is being used and stored.[16]

In terms of *testing* or *analytic activism*, AI could be used to segment contact lists and send more targeted messages to members. Several activists discussed how ChatGPT is useful for managing large email lists and working out how to optimize communications to individual members. One activist explained that AI can 'figure out the best time of day to email someone so that they'll open it [the email]'.[17] AI can also help to discern what issues resonate with people, and personalize content based on whether recipients want lots of emails or see it as spam. Finally, a so far untapped but potential future use of AI is in *facilitating* larger supporter-driven movements. Artificial intelligence could be used to identify which members may be most willing or best at leading new campaigns, for example by looking at past levels of engagement or the size and reach of members' social networks. It could then help connect people with similar or complementary interests and provide a platform for organizing joint actions.

At the time of writing, the use of AI by climate activists is still in its infancy. Many organizations we spoke to were actively considering how to use AI. We are also aware of at least three initiatives whose mission is to support advocacy organizations in integrating AI into their work: *AI Impact Lab*, *Demtech.ai*, and *Daisy Chain*. The founders of these initiatives have extensive experience in digital campaigning and have often been on the forefront of technology innovation in activism. Taren Stinebrickner-Kauffman, founder of *AI Impact Lab*, is the former Executive Director of *SumOfUs*, a digital advocacy organization which targets businesses. Avijit Michael, co-founder of *Demtech.ai*, is former Executive Director of Jhatkaa, an Indian digital advocacy organization. The founders of *DaisyChain* are Nathan Woodhull, a tech developer who has

---

[16] Interview with Amber Macintyre, Tactical Tech, 18 December 2023.

[17] Interview with Indian activist, 16 October 2023.

worked for years with digital advocacy organizations, and Jon Warnow, the former digital lead for 350.org. There are also numerous training opportunities and summits on using AI for organizing and mobilizing.[18]

However, the *transformative* impact of AI on climate activist organizations is not yet apparent. Many NGOs do not have enough high-quality internal data to take advantage of large learning models. As Vivek Katial, Executive Director of Good Data Institute, explained, 'Most NGOs don't have a robust dataset, so they can't use ChatGPT or other LLMs [large language models] to benefit them for better querying, retrieval and understanding of their data. [But] they can still leverage LLMs to support daily tasks such as content generation, emails and even with data analysis.'[19] Even organizations that do have robust data, such as MoveOn, are not yet using AI in transformative ways. As Dave Karpf, professor of political communications, noted, MoveOn are 'trying out AI for the same set of tasks that they were already using machine learning for. They are being mindful of security risks and of AI biases. They are using it to enhance and simplify their existing systems, not to generate original content. *AI, in this case, is functionally a system upgrade*', but not (yet) a transformative game-changer.[20]

In sum, climate advocacy organizations can use digital technologies to scale-up their activities, and to campaign faster and more efficiently. They can also use digital technology to change how they operate and who has decision-making power to shape their campaigns, but importantly not all technology – even AI – has transformative effects on climate activism.

Risks and Challenges of Mobilizing Digitally

An early critique of digital activism was that it would turn supporters into 'clicktivists' (whose main support for social and political causes is via the internet) and lead to 'slacktivism' (superficial engagement with campaigns) that would undermine transformative change (Gladwell, 2010; Merkel, 2017). This argument assumed that activists would mainly use digital technology to mobilize people online (e.g., to post on Twitter or Facebook, or to sign or share online petitions) and that digital activism was based on 'weak ties' rather than 'strong ties' among supporters (Gladwell, 2010). Digital activists would thus channel energy into ineffective tactics, and decision-makers would simply ignore online campaigns. This criticism overlooked how digital technology has enabled the

---

[18] Higher Ground Labs, 'Summit: AI in Campaigning & Organizing', January 30, 2024 www .mobilize.us/highergroundlabs/event/595779/.

[19] Interview with Vivek Katial, 17 January 2023.

[20] Karpf's emphasis. Dave Karpf, 'Bullet Points: A Couple Predictions for AI in 2024', *Substack*, 1 January 2024. https://davekarpf.substack.com/p/bullet-points-a-couple-predictions.

growth of powerful off-line movements – such as Fridays For Future, XR, 350.org described earlier. Many critics did not understand how weak ties networks can diffuse new tactics and support the development of stronger ties.

However, advocates of digital mobilizing and organizing have acknowledged the risk that digitally-oriented organizations may become driven more by the pursuit of 'vanity metrics' than by the quest for transformative change (Silberman and Mahendra, 2015; Rogers, 2018). Vanity metrics include data such as 'list size' (how many people you can email), rates of opening an email, website traffic, or social media followers or likes. These are easy to measure, and present one way to define organizational success. However, they may paint an inaccurate picture of actual support and influence. Many subscribers to digital advocacy organizations, such as MoveOn, do not identify as members of the organization (Fisher, 2019). They have an arm's-length relationship with the organization and may never have met a staff member, let alone built strong relations to other MoveOn members in person. Hence, organizations which primarily campaign online can boast of large memberships, but these members may be inactive. One senior NGO staff member observed that, 'right now, our metrics bias us towards recruiting the most new members, not the right new members' (Silberman and Mahendra, 2015).

Greenpeace's MobLab, one of the early leaders in the field of digital climate organizing, has highlighted the pitfalls of vanity metrics. These include that organizations may make short-sighted decisions, create bad staff incentives, and engage members regularly but without creating change (Holtz et al., 2015). For example, organizations that focus on increasing their list size or number of Facebook followers can easily become distracted from their longer-term mission. The result is 'shallow mobilizing' rather than 'deep organizing' (Han, 2014). Hence, many experts in digital campaigning have encouraged a focus on depth of member engagement rather than breadth. For example, organizations may track how many 'members return for action' (MERA), rather than focus on the total number of members on their list (Silberman and Mahendra, 2015) to measure how successfully they engage their membership.

Another danger is that organizations become overly focused on increasing members' online engagement rather than asking them to take more radical, time-intensive actions that might trigger change. The assumption of many digital campaigners is that new members are more likely to engage with an organization if the 'ask' is easier (e.g., signing an online petition, or sharing a Tweet) and are less likely to be willing to, say, demonstrate outside parliament, chain themselves to a tree, or blockade a fossil fuel company. Digital campaigners often talk of a 'ladder of engagement' – they aim to move members from taking 'easy' actions to more time-intensive, and demanding actions over time. Yet, some digital campaigners view this logic as problematic and argue that many organizations are undervaluing their members' appetite for radical action (Bond and Exley, 2016). Some members may prefer engaging in 'high-bar', time-intensive, and potentially risky actions, including civil disobedience, over 'low-bar' online actions. The success of XR, Sunrise, and Last Generation is precisely due to these organizations asking their members to take high-bar actions, such as blocking major streets by gluing themselves to the road, or throwing tomato sauce onto famous paintings, it is claimed. These direct-action tactics are risky, and many activists have been arrested, but some supporters feel the need to take drastic action given lack of progress on averting catastrophic climate change and are not content with 'easy' online action.

Yet another challenge for climate advocacy organizations relying on digital tools is that they can become dependent on big tech platforms, and vulnerable to changes in their algorithms. Facebook and Twitter (X) are not transparent about their algorithms, and have both changed their algorithms (which determines what material gets seen and goes viral) and rules for who can post on their platforms (Grygiel, 2019).[21] For example, Facebook in 2016 changed its algorithm to prioritize content from users' friends or family over posts from organizations or news platforms (Isaac and Ember, 2016). This made it more difficult for advocacy organizations to reach a broad audience. Facebook has also changed the conditions for posting advertisements about social issues, elections, and politics and requires every social issue organization to be verified (it takes about six weeks to get approved).[22] Many smaller organizations we spoke to found it was much more difficult and sometimes impractical to use Facebook ads to campaign, as Facebook's verification process drained limited time and staff capacity.[23] Tech companies have no interest in being transparent about their algorithms, or to consult the broader public about changes. These are private companies which are driven by 'platform capitalism' and the whims of their leaders, as seen with Musk's takeover of Twitter (X) (Rahman and Thelen, 2019). Further research is needed to explore the role of social media algorithms in shaping organizational strategies and public perceptions of climate politics.

A final, fundamental, risk of any digital organization or campaign is the security of members' data. A specific problem for activists using digital platforms is that people are encouraged to regularly share their political views online. This is particularly a risk in authoritarian regimes, where civic action is often highly restricted if not illegal (Gohdes, 2020). In addition, many climate organizations have large databases with potentially sensitive data about their

---

[21] It is easy to forget that Facebook, for example, was initially created for students and only people with an .edu email could register.

[22] Facebook, 'Business Help Center About ads about social issues, elections or politics', www .facebook.com/business/help/167836590566506?id=288762101909005.

[23] Interview with Joshua Low, 23 September 2023.

members – their names, addresses, and credit card details. If these organizations are hacked, and the data falls into the wrong hands, it could be used to attack or 'dox' them. Tech companies are also known to hand over private data to governments, hence even private messages (e.g., on WhatsApp) can be risky. Scholar Lucy Bernholz has championed the motto, 'don't collect, what you can't protect' to ensure that organizations do not collect potentially sensitive information (Schulz, 2018). In the final section of this *Elements*, we discuss further how digital tools are used to repress climate activists.

Conclusions and Questions for Further Research

Social media has enabled climate organizations to increase their scale and coordinate global movements for climate action. Organizations like 350.org, Fridays For Future, Extinction Rebellion, and Last Generation have used social media and email to mobilise thousands to march in the streets and take direct actions such as blockading banks and highways. Much of the impact of these movements has been due to their off-line tactics. However, the fact they were able to scale up so quickly, and diffuse tactics globally, is thanks to digital technologies.

We have argued that scholars should focus not only on how technology has changed the *scale* of organizing by climate activists, but also on *how* organizations use technology to campaign and mobilize their members in novel ways. Rather than focusing on what platforms activists use, we should be asking: are advocacy organizations using digital platforms to *broadcast* their messages, *converse* with their members, *test* their campaigns, and/or *facilitate* global movements (Hall et al., 2020)? These different strategies are based on different theories of change (inside versus outside, mobilizing versus organizing, etc.). Climate activists should reflect further on how they use digital technologies and digital analytics to support their work. Legacy organizations which have conventionally relied on broadcasting strategies to communicate may be reluctant to hand over power to their members to decide campaigning actions and tactics. Yet, by distributing power to members and volunteers they may gain greater influence (Han, 2014; Bond and Exley, 2016). Technology is not simply a communication tool but can also transform power relations within organizations and facilitate innovation.

Scholars should also look at when climate advocacy organizations have successfully harnessed digital technologies to mobilize and organize off-line. The impact of digital technology is not limited to the digital world as we have shown in this section. Scholars could further look at varying levels of digital campaigning expertise: how much do different climate advocacy organizations invest in staff with digital expertise; and how are data analytics integrated into strategic decision-making? What knowledge do activist organizations have of how algorithms work and how they can amplify their messages? How do different levels of digital campaigning expertise influence campaign reach and success? Scholars might also use population ecology approaches to examine if and how groups with digital mobilizing expertise can complement advocacy organizations who have off-line organizing expertise, thus boosting the overall impact of climate advocacy (Bush and Hadden, 2019; Eilstrup-Sangiovanni, 2019). Finally, we also need more comparative work on the challenges climate advocacy organizations face in authoritarian regimes compared with democratic regimes when campaigning online (Gohdes, 2024) and how groups in the Global South may be leap-frogging organizations in the Global North (e.g., through use of peer-to-peer messaging). There are far fewer studies of digital climate movements in the Global South, with the exception of China which is comparatively well studied (Baran and Stoltenberg 2023:462).

Overall, this section encourages practitioners and scholars to reflect on how digital technology is enabling fast, broad, global climate movements to form and may also be changing the balance of power between legacy NGOs (such as Friends for the Earth and Greenpeace) and newer climate organizations (such as FFF, XR, and Sunrise). These newcomers, we suggest, are strong, not because they use digital technology, but because of how they use it to distribute power to members.

2 Monitoring and Enforcement

In 2015, the Ecuador Waorani people embarked on a mission to map the boundaries of native communities in the Pastaza region in the heart of the Ecuadorian Amazon – an area zoned by the government for oil exploration. The mapping, which relied on geospatial technologies such as satellite images, GIS, Global Positioning System (GPS), and digital photography, covered 1,800 square kilometres, and identified 1,832 routes consisting of rivers, streams, hunting trails, and paths that served to connect their communities. It also recorded close to 10,000 specific geo-referenced points, including sensitive ecosystems and areas for timber, fishing, and hunting. Mapping was done in small teams under the guidance of 'Pikenanis' (local 'wise guides' – typically community elders) who worked with local technicians trained in using GPS, digital recording, and various software to record findings (Aguilar, 2018). Four years later, in 2019, these digital maps formed part of the evidentiary basis for a successful court case against the Ecuadorian Government. The judge's landmark ruling held that permission to extract oil on Waorani lands had been granted without due consultation with ancestral landowners who could demonstrate a longstanding relationship to the land and ruled in favour of protecting a large swathe of the Waorani rainforest.[24]

---

[24] https://news.mongabay.com/2018/06/ecuador-waorani-people-map-their-rainforest-to-save-it/.

The Waorani example illustrates how digitally based monitoring can empower indigenous communities to fight against climate change. Indigenous peoples' stewardship of ancestral forested lands is increasingly recognized as critical in efforts to combat climate devastation (Chapin et al., 2005; Ijjasz-Vasquez and Betancourt, 2023). Indigenous territories encompass 22 percent of the world's land surface and hold 80 percent of the planet's biodiversity (World Bank, 2008). Yet, many indigenous peoples are challenged by governments' failure to recognize their land rights. Since the 1970s, native communities have used geospatial technologies, including satellite images, GIS, and GPS, to protect tribal lands from cattle ranching, logging, and oil exploitation (Wickens and Louis, 2008). However, recent decades have seen a sharp rise in initiatives focused on unlocking the potential of indigenous digital mapping, as the shift to online mapping platforms has created opportunities for more people than ever before to create, shape, and share digital maps which can support political advocacy and legal action (de Roy, 2021;[25] Ijjasz-Vasquez and Betancourt, 2023; Gutiérrez, 2018; Cifuentes, 2020). An apt example of this trend is Digital Democracy, a small NGO that specializes in working with communities in the Amazon to harness technology to defend their land rights.[26] Digital Democracy has developed an open-source software tool, 'MAPEO', that allows native peoples to collect and store data about their lands and share these with others in a private closed network. Importantly, MAPEO can be customized to more than fifty local language settings, helping to overcome the cultural and language barriers faced by many indigenous groups online. Another example is Indigenous Mapping Collective – a virtual platform developed in partnership with Google Maps and Google Earth Outreach to allow indigenous groups to access digital mapping resources.[27]

The Waroani's successful use of geospatial tools to defend their land rights is not unique. In Paraguay, a game-changer for the roughly 250,000 indigenous people living in the Gran Chaco has been the Tierras Indígenas – an online platform which combines data from Global Forest Watch[28] and software from the World Resources Institute's MapBuilder[29] with information collected by local indigenous groups to map the geographic footprint and establish the legal status of their natural resources (Ijjasz-Vasquez and Betancourt, 2023). In Peru, the Cadasta Foundation which supports indigenous communities in 45 countries with GIS technologies and training to secure inclusive land rights for climate

---

[25] Steven de Roy is Founder of the Indigenous Mapping Collective.
[26] www.digital-democracy.org/mapeo.   [27] www.indigenousmaps.com/.
[28] https://data.globalforestwatch.org/datasets/3d668cf0fbcb415bba1ec00bc6263877_5/explore?
location=-25.001563%2C-60.133738%2C4.78
[29] https://mapbuilder.wri.org

sustainability,[30] has trained indigenous mappers from the Awajún and Asháninka communities to deploy digital technology to safeguard 11,000 hectares of rainforest, sequestering 5 million tonnes of carbon (Ijjasz-Vasquez and Betancourt, 2023). What these examples have in common is that they involve indigenous communities combining local knowledge and tradition with modern technology to challenge governments to recognize their stewardship over critical climate assets.

Building from examples like these, this section considers how new technologies contribute to challenging the monopoly of the state over critical information and data infrastructure (Beraldo and Milan, 2019:2). For decades, if not centuries, scientists and environmental policy experts have collected and analysed large-scale data to monitor climatic and other environmental changes on a local and global scale. The launch of the first Earth-observing satellite program (Landsat) by NASA in 1972[31] generated a wealth of new remote sensing data for use by government agencies, major research institutions, and private companies (Baker and Williamson, 2006). This data has been used to define climate targets, develop regulations, and address environmental harms affecting local communities (Engine Room, 2022:61). The production and use of remote-sensing environmental data, however, was predominantly state controlled. In contrast, the recent upsurge in commercial remote-sensing technology and publicly available satellite imagery, combined with the ability to crowdsource data using online platforms or simple smartphone devices, has increased both the amount and kinds of information that can be collected by non-state actors. Thanks to such technologies, many grassroots groups, NGOs, and indigenous communities today have access to sophisticated data-gathering and data-analysis tools that were once the preserve of governments and public scientific agencies (Aday and Livingston, 2009; Rothe and Shim, 2018; Kazansky et al., 2022).

Noting the widening access to data, this section explores to what extent, and how, digital technologies such as remote sensing (e.g., satellite, drones and GPS), Geographic Information Systems (GIS), and 'big data' analytics can serve to level the playing field between non-state actors and states (as well as corporate actors) by facilitating new forms of monitoring and enforcement. Data – or more simply 'information' – has long been viewed as a critical resource for non-state activists. As Keck and Sikkink (1998) observed a quarter of a century ago, 'at the core of NGO activity is the production, exchange and strategic use of testimonial, statistical and scientific information … to document problems and suggest possible solutions' (p. x). As Beraldo and Milan

---

[30] https://cadasta.org/landforclimate/; and https://landportal.org/node/116598.
[31] www.usgs.gov/faqs/what-landsat-satellite-program-and-why-it-important.

(2019:2) note, such activities take on new dimensions in the age of data-driven politics, as publicly available data can be repurposed as a 'new currency' in the relationship between the state and activists. For example, activists can today analyse vast amounts of 'open data' made available online by public administrations to find evidence to support their claims towards the state, or to hold private corporations accountable for environmental harms. Such monitoring and enforcement capacities are crucial for climate action, given that many governments and corporations have committed to taking action to keep global average temperature increases below 1.5–2.0 degrees yet fail to deliver on these promises.

In addition to focusing on a wider range of technologies and uses beyond social media and email, this section also brings into view a broader range of non-state climate actors, including research institutes based in universities, transnational communities of scientists, and non-profit start-ups that support NGOs in developing technologies for monitoring and enforcement. We begin by illustrating how the diffusion of affordable remote-sensing technology like satellite-based GPS, GIS, satellite imagery, digital sensors, drones, and more, enables climate activists to gather independent information on issues like deforestation, changes in land use, and GHG emissions. Next, we consider how open-source data and 'citizen science' enable activists to monitor complex environmental factors – from country-level emissions to corporate greenwashing. Finally, we illustrate how remote sensing and advanced data-analytics underpin activists' efforts to ensure enforcement of climate policies and regulations independently of states, through either direct action or litigation. As in other sections, we do not strive to provide a representative picture of 'typical' uses of remote-sensing technologies and open-source data by NGOs. Rather we illustrate how such technologies are leading to new ways of fighting for climate action through digital monitoring and enforcement.

Private Eyes in the Sky: Monitoring Global Forests

The upsurge in climate monitoring by non-state actors is perhaps nowhere more visible than in activists' growing use of satellite and other remote-sensing tools to observe and document large-scale environmental developments such as deforestation. Unlike aerial photography, satellite imaging does not depend on prior approval from governments or private landowners. Instead, satellites can capture images of large areas – whether terrestrial, oceanic, or atmospheric – almost anywhere in the world on a routine basis (Baker and Williamson, 2006). As a result, even small non-profits can increasingly monitor global forests and oceans, measure the melting of glaciers, catalogue the loss of natural habitats, and track GHG emissions by major industries. While the high cost of 'bespoke' satellite imagery remains a barrier for many climate organizations, many smaller organizations rely on publicly available satellite imagery from sources like Google Earth or use crowd-sourced data from close-range sensing mechanisms such as hand-held camera devices, Internet of Things (IoT) sensors, and unmanned aerial vehicles (UaVs), which have made remote data-collection doable on a smaller budget (Engine Room, 2023).

A prominent platform harnessing satellite technology to fight climate change is Global Forest Watch. Launched in 2013 by the World Resources Institute in cooperation with other non-profit, private, and public institutions, Global Forest Watch is an open-source web application that uses satellite imagery to monitor global forests in (close to) real time.[32] While it provides data to public authorities and international organizations across the world, Global Forest Watch is also a crucial resource for non-state climate activists. The platform allows users anywhere to scroll through data, zoom in on any location and visually combine a wide range of social, economic, and environmental databases connected on the platform. Climate organizations around the world can use this data to inform their campaigns, and to support the development of other monitoring platforms as seen in the earlier example of the Tierras Indígenas platform for mapping indigenous territories.

Platforms similar to Global Forest Watch have been launched on a smaller scale in countries with major tropical forests, including Indonesia, the Congo, and Brazil (Aday and Livingston, 2009). Importantly, these platforms combine satellite imagery with data from local forest groups to monitor and publicize activities linked to deforestation.[33] An illustrative example is the Indonesian Independent Forest Monitoring network, Jaringan Pemantau Independen Kehutanan, which connects local NGOs and indigenous and community groups trained in how to conduct forest monitoring activities, using a blend of digital and analogue methods.[34] Another illustration is Project Canopy, a small non-profit that uses machine learning to parse satellite imagery and identify illegal logging in the Congo Basin Rainforest and then sends alerts to local policymakers and conservation groups.[35] These are just two examples among hundreds of similar projects around the world which use remote sensing to monitor and protect forest

---

[32] www.wri.org.
[33] https://capacity4dev.europa.eu/discussions/new-report-highlights-importance-indonesian-independent-forest-monitorings-contribution-svlk_en.
[34] https://capacity4dev.europa.eu/discussions/new-report-highlights-importance-indonesian-independent-forest-monitorings-contribution-svlk_en.
[35] www.projectcanopy.org/#What-We-Do.

resources at both local, national and international level. Importantly, remote sensing is not limited to satellite imagery. Use of low-cost drones is also growing, especially in Latin America, Indonesia, and India (Global Information Society Watch, 2020; Paneque-Galvéz et al., 2017). For example, indigenous communities in the Amazon rainforest use drones and camera traps to monitor ancestral territories for invaders, including illegal loggers, and to estimate forest carbon.

The growing availability and declining cost of remote sensing have the potential to alter power dynamics between state and non-state actors 'by allowing activists to overcome the problem of political "scale differentials"' (Aday and Livingston, 2009:515; Rothe and Shim, 2018). As Aday and Livingston note, before the tech boom, the range of opinions and parameters of discussion on climate change in the press tended to be institutionally based and elite-driven. Today, the declining cost of orbit satellite and other remote-sensing tools mean NGOs can credibly question official data on deforestation, land use, GHG emissions, and atmospheric changes. In turn, this can force governments to address issues they might have preferred to keep quiet (ibid 514.; Rothe and Shim, 2018:420). Yet, there are also important limitations to keep in mind. First, it is still relatively expensive for NGOs to acquire real-time satellite imagery of a high spatial or spectral resolution.[36] Thus, access to useful satellite data remains uneven. Second, accurately interpreting satellite imagery requires a high degree of technical expertise. Many activists lack the necessary training and experience in remote sensing to draw clear inferences or to produce credible evidence equivalent to that of geospatial experts in state agencies, industry, and academia (Baker and Williamson, 2006:12). Third, to establish causal links and triangulate evidence, it is often insufficient to rely on satellite data, as activists may need to 'ground' observations by taking close-up photos or video, and by gathering oral histories and narratives to fully understand the context and extent of threat to the specific resources that matter most to local stakeholders. These limitations notwithstanding, the extension of remote sensing technologies and skills to marginalized communities can often allow them to gather and establish local ownership over new types of data which can bolster their claims against governments and corporations.

Tracing Emissions at Home and Abroad

Remote sensing technology not only can help climate activists to defend ancestral lands or protect global forests against illegal logging; against the backdrop of growing public pledges by states and corporations, technology has also opened

---

[36] Notes from fieldwork on board Sea Shepherds' ship in Gabon April 30, 2023. Interview with crew by Teale P. Bondaroff.

new possibilities for activists to effectively compute and monitor GHG emissions. Take the example of ClimateTRACE. ClimateTRACE is a coalition of hundreds of NGOs, tech companies, and universities[37] which offers an independent platform to track greenhouse gas emissions. The initiative was born in 2019 when two smaller NGOs, WattTime and TransitionZero, received a grant from Google.org to use satellite data to monitor emissions from power plants. At the time of writing, some five years later, ClimateTRACE combines data from more than 300 satellites and more than 11,000 sensors with AI and machine learning models to pinpoint individual sources of GHG emissions, making its data publicly and freely available. What is perhaps most striking is how quickly the initiative has been able to scale up. The first global emissions inventory was released in September 2021 providing *country-level* data on emissions. By November 2022, ClimateTRACE released its first *facility-level* inventory, covering more than 72,000 of the biggest known individual sources of GHG emissions across two dozen industries. By December 2023, the inventory covered more than 352 million individual facilities.[38] The data is freely available on ClimateTRACE's online platform, allowing users to browse individual assets via a map function, download and compare full country- or sector-level datasets, and access asset-level ownership data. As such it opens new opportunities for climate action and for holding governments and corporate actors accountable for breaches of regulations or empty green pledges on an unprecedented scale.

Indeed, a major benefit of remote sensing technology is that it can enable activists to reveal discrepancies between official emission reporting and actual behaviour. Derived directly from satellites and other remote sensing methods, ClimateTRACE's inventories capture emissions that may go unreported in traditional official inventories based primarily on self-reporting. They also capture emissions by actors such as national militaries that do not have to fully report their emissions under Kyoto Protocol reporting requirements. Analysts have found that the *majority* of corporate emissions data captured in ClimateTRACE's inventories are not included in self-reported environmental, social and governance (ESG) databases.[39] Many countries also fail to report their emissions accurately, despite obligations to do so under the Paris Agreement. Digital monitoring platforms like ClimateTrace offers a mechanism for holding them to account.

---

[37] Coalition members include, among others, Blue Sky Analytics, CarbonPlan, Earthrise Alliance, Hudson Carbon, Hypervine, Johns Hopkins Applied Physics Laboratory, OceanMind, RMI, TransitionZero, WattTime. https://climatetrace.org.

[38] https://climatetrace.org/news/climate-trace-unveils-open-emissions-database-of-more-than.

[39] ESG data refers to environmental, social and governance data that companies disclose to account for their sustainability and ethical practises.

In addition to revealing discrepancies between official reporting and actual emissions, ClimateTRACE, (and other similar platforms), can boost climate advocacy by pinpointing specific areas for intervention. For example, the 2023 ClimateTRACE inventory revealed that the biggest growth in GHG emissions since 2021 was caused by electricity production in China and India and gas production in the US. It also showed rising emissions of methane despite more than one hundred countries signing up to a pledge to reduce the gas during COP26 in 2021.[40] Such specific, and prompt, data makes it easier and faster to design and mobilize campaigns to hold governments accountable to their words. Besides documenting aggregate emissions, the inventory allows users to identify specific sources of emissions, down to the level of individual industrial facilities, thereby facilitating campaigns to 'name and shame' specific industries or corporations or, as we discuss later, to initiate lawsuits against individual offenders. Finally, like other large-scale monitoring platforms such as Global Forest Watch, ClimateTRACE's inventories can be integrated with data from other platforms to support a wide range of climate activism. To illustrate, the non-profit The Joint Impact Model (JIM) integrates ClimateTRACE's data with data from financial institutions to estimate GHG emissions and climate impacts of major investment portfolios.[41] Challenges and risks notwithstanding (we discuss these further at the end of this section), these examples show how remote-sensing technology can help change the balance of power between climate activists, states, and corporations and ensure greater public accountability of governments and corporate actors.

'Connecting the Dots': Using Geospatial Information to Hold Dirty Corporations to Account

Moving on from predominantly space-based and aerial monitoring of GHG emissions to technologies closer to the ground, we return to another tool in the toolbox of contemporary climate activists: advanced Geographic Information Systems (GIS) which enable real-time monitoring of climate assets, as well as detailed investigations of specific environmental harms. Simply put, GIS provides a framework for collecting multiple types of data, from geographical and topological data gathered by satellite to statistical and qualitative information entailed in local databases or news archives and integrating these into a combined system that allows for advanced analysis and visualization (Engine Room, 2023). While 'old-style' datasets might record location-based information like postcodes, city names, population density, and so on, GIS systems include a further range of geographic attributes, for example, the extent of droughts, wildfires, biomass changes, or flooding in a location. These

---

[40] See, for example, https://climatepolicyradar.org/.   [41] www.jointimpactmodel.org.

separate bits of information can then be layered on top of each other and visualized in sophisticated virtual maps or 3D models, using specialist software (Engine Room, 2023).

Geographic Information Systems provide another illustration of how digital technologies can change power balances between climate activists and target actors like states and corporations. Early GIS was mainly used by public institutions because they required expensive, high-end hardware and software. Yet today, GIS can be run on standard computers, or even smartphones, making the technology more affordable and accessible. A great example of how GIS facilitates climate-focused monitoring and enforcement is Global Witness' investigation linking multinational beef traders in the Brazilian rainforest to deforestation (Global Witness, 2020).[42] To expose the link, Global Witness obtained all cattle transport permits granted by Brazil's Federal Government between 2017 and 2019 from the website of the Sanitary Agency of the state of Pará. Activists used these publicly available documents to determine which ranches supplied cattle to the three largest Brazilian beef traders: JBS, Marfrig, and Minerva. Next, activists downloaded the physical boundaries of those ranches using online data from the environmental registry of Pará state and matched these with the transport permits. These data were then overlaid with Landsat and Sentinel satellite imagery to check whether deforestation had occurred within the ranches, using a visualization tool called TERRAS. Finally, activists used Pará state and federal databases of deforestation permits[43] to check whether recorded deforestation in the ranches was legal or illegal. This analysis revealed that, in 2019 alone, the beef company JBS purchased cattle from more than 200 ranches containing 17,000 hectares of rainforest deforestation, none of which had the required deforestation permits. Nevertheless, a recent official audit of farms in the JBS supply chain had found only one in six farms to be engaged in deforestation to make space for cattle grazing, and none were deemed to be non-compliant with Brazil's Forest Code. Thanks to Global Witness' subsequent work with local prosecutors, JBS was fined nearly $1 million and forced to drop many of its suppliers (Global Witness, 2020).

Global Witness' campaign to expose the climate cost of Brazil's beef industry illustrates how integrating different sources of data can enable climate activists to conduct 'forensic' digital analysis to evidence unlawful behavior and bring culprits to justice. Thanks to technological innovation, NGOs today have unprecedented capacity to obtain and analyze evidence in ways that can be useful to prosecution efforts (Langer and Eason, 2019:764; Koenig, 2017;

---

[42] Global Witness, 'About Us', www.globalwitness.org/en/about-us/ (last accessed 6 May 2023).

[43] Brazil's Forest Code requires rural producers that want to deforest in their property to obtain a permit.

Eilstrup-Sangiovanni and Sharman, 2022).[44] Although investigative journalism, exposés, and independent investigations leading to lawsuits against polluters are nothing new, digital tools supercharge such strategies by lowering the cost of evidence gathering, and by enabling activists to produce a level of proof that would have been difficult in the pre-digital age. In this context, 'open-source intelligence', finding evidentiary needles in the haystack of free online information, has become key (Higgins, 2021:182). Without access to open data and digital analytics it would have been difficult for Global Witness to establish a clear link between specific beef ranches and deforestation, or to gather sufficient evidence to bring a successful court case. However, documenting wrongdoing and linking crimes to specific perpetrators is only half the game. As we discuss later in this section, although many NGOs and indigenous groups have taken to the courts to press their claims against governments or corporations, not all cases deliver favourable outcomes. Legal processes are often subject to long delays as defendants appeal judgments. Those found guilty often ignore fines or other injunctions and fail to change their behavior. Hence the sharp uptick in climate-related litigation, while promising in terms of raising awareness, and possibly helping to deter some specific climate harms, may often fail to get at the underlying problems driving climate devastation.

Engaging the Public through Crowdsourcing and Citizen Science

Many of the examples featured so far in this section involve NGOs, digital experts, and researchers working either *on behalf* of local communities or working to empower local actors by lending expertise or providing digital training and resources. But digital technologies also enable more direct participation and self-directed action by non-experts. Similar to the decentralized forms of organization discussed in Section 1 whereby climate groups hand agenda control and initiative to supporters to design their own actions, technologies for crowdsourcing evidence and 'citizen science' platforms open opportunities for the wider public to become directly engaged in monitoring, data-gathering, and scientific testing – or even to launch their own monitoring and enforcement projects.

Crowdsourcing involves obtaining information or other input into a task or project by enlisting the services of large numbers of people or inanimate entities, typically via the internet (Howe, 2006). Whereas satellite imagery and other geospatial data must be purposefully purchased, downloaded, recorded, and integrated by trained analysts, machine-learning tools allows analysts to combine data that is routinely collected by public and private

---

[44] Eyewitness interview, 2020. See also www.eyewitnessproject.org.

sensors, amateur weather stations, and other smart devices such as smart-phones connected via the internet or other systems (Mueller et al., 2015:3185; Milan, 2018; Beraldo and Milan, 2019). Sometimes referred to as 'citizen sensing' such approaches are often complementary to more traditional remote sensing approaches.

Whereas crowdsourcing may be inanimate; involving the acquisition and repurposing of data from a range of sensors connected via the internet (say domestic smart-meters or local weather stations), many crowdsourcing projects rely on active participation by volunteers in data creation and analysis (Muller et al., 2015). One venue for crowdsourcing climate research is Zooniverse[45] which, at the time of writing (February 2024), hosts 98 projects engaging volunteers in projects ranging from researching the effects of water-level changes due to climate change on the ecology of Great Lakes' coastal wetlands, to tracking the progress of forest restoration efforts around the world ('RESTOR'), or recovering the data of Argentine historic weather records to improve understandings of climate change (Meteororum ad Extremum Terrae, MET).

How do ordinary citizens participate in such projects? To reduce the time and costs of monitoring forest restoration, scientists at RESTOR have developed an AI model that identifies trees in photos taken by drones. To train the model, they feed it thousands of images with pre-labelled trees. Volunteers are asked to tag trees to validate the AI training data. Members of the public are thus invited to take active part in climate research whose objectives and methods have been centrally defined by a team of scientists and professional staff and that, in the examples of both RESTOR and MET, is supported by large organizations like UNEP, WWF, Google, and the World Meteorological Organization. However, there are also examples of citizen science taking more decentralized forms, mirroring the kinds of supporter-led or member-led climate action discussed in Section 1. For example, specialized apps such as 'Epicollect'[46] offer a free mobile data-gathering platform that allows groups to set up projects to crowd-source information, while the Association for Advancing Participatory Sciences[47] helps scientists, practitioners, and activists across the world to build and operate their own citizen science projects.

In addition to reducing data collection costs, crowdsourcing can expand the resources and capacity of climate activist groups. Climateprediction.net (CPND), a volunteer computing and climate modelling project based at the University of Oxford's e-Research Centre, has since 2003 relied on volunteers to run climate modelling simulations on their home computers. The data

---

[45] www.zooniverse.org.   [46] https://five.epicollect.net/.   [47] https://participatorysciences.org/.

generated through distributed computing simulations is sent back to CPND and incorporated into aggregate models.[48] Climate models are generally large and extremely resource-intensive to run. By relying on distributed computing power from thousands of volunteers, CPDN can run climate experiments that would be too computationally demanding for even supercomputers.[49] Scientists interested in running simulations can fill in a collaboration enquiry form on the CPDN website.

Crowdsourcing and citizen science hold significant potential to enhance the effectiveness and lower the costs to climate activists of collecting and analyzing large amounts of environmental data (Willis et al., 2017; Kirilenko et al., 2017). Engaging citizens in data collection may also increase public awareness and knowledge, thereby empowering communities to take more effective climate action. The public often receive conflicting claims regarding the causes and effects of climate change via the media, from scientists, and from political representatives. Yet, as 'ordinary' people become active co-producers of environmental information and scientific data, power relations begin to change. Thanks to bottom-up, citizen-powered data collection and volunteer monitoring, climate activists can increasingly challenge dominant political and scientific narratives with credible data that has a substantial degree of public buy-in.

There are, of course, also limitations to citizen sensing and citizen science. As Glicksman and co-authors (2016) warn, data generated by individuals and local communities may be 'self-selected with unsure representativeness'. Another potential pitfall is the growing commercial interest in remote sensing. As many examples in this section illustrate, platforms for digitally distributed climate monitoring and climate modelling, whilst powered by volunteer efforts, are often designed on a pro bono basis by employees of tech start-ups or funded by generous grants from large tech companies like Google.org. Despite their nonprofit basis, profit motives (via potential commercial applications) may lurk in the background of some such projects. Indeed, projects such as RESTOR may be seen to raise ethical concerns about whether volunteers are essentially being used as free labour to gather data on which to train AI models that may also have commercial applications. As for-profit motives enter, public interests may give way to private ones.

Leveraging Open Data to Push for Political Change

Just as technology can help climate activists to collect independent data at greater speed and lower cost than ever before, it also provides access to a wealth of already existing data which can assist monitoring and enforcement. As we have

---

[48] Climateprediction.net (last accessed 1 July 2024).   [49] Climateprediction.net.

seen, data about our physical and social environment is continuously being generated by various technologies – from satellites and aerial sensors to surveillance cameras and personal mobile devices. Much of this data (such as text, video and audio files, links and tags resulting from online distribution, and meta-data generated by interactions on social networking platforms) constitutes what Cukier and Mayer-Schoenberger (2013) call 'data in the wild'. That is, it is generated randomly (Milan, 2018). However, this data can still be amassed and put to specific uses thanks to powerful processors, smart algorithms, and advanced software (Cukier and Mayer-Schoenberger, 2013; Milan, 2018). Meanwhile, vast amounts of data are purposely put in the public sphere. 'Open data' refers to data that is freely available online for anyone to access, download and share. 'Open government data' (OGD), in turn, denotes the practice of making data from government agencies freely available on digital platforms to enhance transparency and encourage public participation.[50] Both kinds of data create opportunities for activists to monitor target actors and, in some cases, to collect evidence to launch legal actions.

An example of how open data can enable new forms of climate activism is the Climate Action Tracker. The project, which was launched in 2009 by two non-profit organizations, Climate Analytics and the New Climate Institute, is dedicated to tracking government climate action using open-source data. Climate Action Tracker collates existing datasets (including GHG data submitted by governments to the UNFCCC, national inventory reports and databases, and data published by the International Energy Agency and select scientific institutions) to produce a rating of which countries and sectors are taking sufficient action to align with the Paris Agreement's goal of limiting global warming to $1.5°C$.[51] The initiative evaluates climate mitigation targets, pledges, policies, and actions for 39 major countries and the EU (covering around 85 percent of global emissions) and serves as a crucial source of information for many other climate activist organizations. A similar initiative, the Global Footprint Network, collects and publicizes data on the ecological footprint and biocapacity of more than 200 countries and regions around the world, based on UN or UN affiliated datasets like those published by the Food and Agriculture Organization, the UN's Commodity Trade Statistics Database, and the International Energy Agency, along with studies in peer-reviewed scientific journals, all of which are integrated to provide a single, country-specific

---

[50] Official OGD programs have been launched in many countries which require administrative agencies within a government to make their data discoverable, available, and downloadable through dedicated internet portals. OECD, www.oecd.org/gov/digital-government/open-government-data .htm; OpenGovernmentPartnership, www.opengovpartnership.org/our-members; Dawes et al., 2016).

[51] Climate Action Tracker, https://climateactiontracker.org/ (last accessed 24 April 2023).

measure of overall ecological resource use.[52] As well as serving a crucial monitoring function, these and similar platforms for integrating publicly available climate data can also open opportunities for activists to engage directly with policymakers (especially with democratic governments) about where and how to intervene to improve public policies.

Another example of how Open Government Data can empower climate activism is activists' use of pollutant registers such as the European Pollutant Release & Transfer Register (E-PRTR) which records data on pollutant releases and off-site waste transfers from 50,000 industrial facilities in the EU and affiliated countries. On its face, E-PRTR is simply a public pollution accounting system that requires large manufacturing firms or industrial facilities to report their annual emissions. Emissions are self-reported and not closely monitored by public authorities (Fung and O'Rourke, 2000). Once available online, however, this data can be used for purposes well beyond those imagined by legislators (Wotzka, 2022). For example, Greenpeace and Friends of the Earth have used E-PRTR data to estimate premature deaths due to hazardous substances released from European coal plants, and to measure direct and indirect contributions to global warming from industrial livestock farming.[53] Despite likely under-reporting emissions, online public registers can thus serve as powerful campaigning tools.

Combatting Corporate Greenwashing

Closely related to monitoring the climate performance of governments and public agencies through open data analytics are efforts to combat corporate greenwashing. As climate change has become more politically salient, there has been a surge in greenwashing, or 'climate washing', whereby businesses exaggerate their progress towards 'net-zero' pledges and other self-professed climate goals through misleading advertisement (CAAD, 2023; Setzer and Higham, 2023; Liu et al., 2023). False claims and empty pledges mislead consumers, investors, and regulators, and hinder effective policy interventions. The problem was highlighted during COP27 in 2022, as the United Nation's High-Level Expert Group on the Net-Zero Emissions Commitments of Non-State Entities issued a report entitled 'Integrity Matters' which called for stronger criteria for net-zero pledges by corporations.[54] The report noted that 'there is currently no international verification system for net-zero pledges'

---

[52] https://data.footprintnetwork.org/#/abouttheData; *Footprint Data Foundation.* www.FoDaFo.org.

[53] Greenpeace (2013). Studie: Tod aus dem Schlot. www.greenpeace.de/publikationen/studie-tod-schlot.

[54] www.un.org/sg/en/content/sg/statement/2022-11-08/secretary-generals-remarks-launch-of-report-of-high-level-expert-group-net-zero-commitments-delivered.

meaning that citizens, consumers, and investors are 'largely operating in the dark' when seeking to make climate-friendly choices.

A necessary basis for challenging corporate greenwashing is to obtain clear and reliable data to allow governments, consumers, investors, and civil society actors to verify companies' claims (UN High Level Expert Group on Net-Zero Emissions, 2022:41). As Kazansky and co-authors (2022) observe, 'greenwashing thrives in an environment of information asymmetry'. In this context, the ability to integrate complex data from corporate business plans, social media, advertisement, and online emissions databases can enable climate activists to more effectively pinpoint and document, greenwashing. For example, Greenpeace has used open-source data from Corporate Social Responsibility (CSR) reports and submissions to public pollutant registries to monitor big tech firms and publish report cards on how green they are.[55] Other groups, like Global Witness, have compared corporate media communications with emissions data to check the green claims by corporations.[56]

The methodologies used in open-data climate indexes such as the Climate Action Tracker and Global Footprint Alliance may be open to criticism, and the public data on which they draw may be woefully incomplete. Nevertheless, the growing availability of open data constitutes an important resource for climate activists to monitor the behaviors of governments and corporate actors, and to push for changes to policies and regulations. Still, there are limitations to open-data activism. First, although calling out greenwashing may help to put pressure on corporations and policymakers, it does not by itself achieve much in the absence of stronger regulation of net-zero and other climate pledges. Second, and more broadly, despite the emphasis on transparency, publicity, and citizen engagement, users of open data – and open government data specifically – are typically not 'ordinary citizens' but technologically skilled data analysts or application developers who are able to navigate data in highly technical formats (Dawes et al., 2016). These actors may focus mainly on problems that affect them (or their funders), or that lend themselves well to quantification and visualization in particular digital formats. Also, while expert analysts may use open data to create applications like the Climate Action Tracker free of charge, they can also use open data as a basis for developing revenue-generating applications that are off limits for many stakeholders (Dawes et al., 2016). As we discuss in more depth in Section 5, this highlights entrenched problems of unequal access and uneven distribution of technical capacities to make effective use of new technologies.

---

[55] www.greenpeace.org/usa/reports/greener-electronics-2017/.
[56] https://www.globalwitness.org/en/campaigns/greenwashing/.

Enforcement

Rounding off this section, we consider how new technologies can help climate activists contribute directly to enforcement of environmental laws and regulations. The pervasiveness of corporate climate washing highlights the limitations of relying exclusively on monitoring and 'naming and shaming' to change behavior. Calling out bad behavior may be effective in some cases, but to really change incentives there often needs to be a possibility of imposing sanctions for continued wrongdoing. As well as improving monitoring, many of the technological innovations discussed in this section also increase the capacity of climate activists to assist in enforcing environmental laws and regulations through independent investigation and litigation (Eilstrup-Sangiovanni and Sharman, 2022). For example, recent scholarship documents how the growing sophistication and declining costs of internet-connected devices that can be used for audio, photo, and video recording allow climate activists to document environmental abuses and submit court-ready evidence to public prosecutors with growing ease (Langer and Eason, 2019:794; Eilstrup-Sangiovanni and Sharman 2021, 2022). Digital technologies also play a growing role in enabling activists to pursue enforcement of climate laws independently of police and public prosecutors by bringing legal cases before national, regional, and international courts (Eilstrup-Sangiovanni and Sharman, 2022).

The ways technology facilitates law enforcement are both similar and different to how it underpins more general monitoring. Fundamentally, intervening against environmental crime requires detailed and reliable knowledge of where and when it occurs. By leveraging new surveillance technologies, NGOs are arming themselves (and/or public authorities) with actionable evidence on where and when to intervene. Some groups use remote sensing and supporting tools such as GIS as a basis for direct interventions against lawbreakers (Eilstrup-Sangiovanni and Sharman 2022). For example, maritime groups like Sea Shepherd have used satellite imagery and GPS to identify, locate, and physically block illegal fishing boats from causing damage to protected marine areas (Eilstrup-Sangiovanni and Sharman 2022). Publicly available satellite surveillance platforms like Global Forest Watch, which provide real-time information about activities in global forests, also provide a basis for direct interventions against illegal logging.

Technology is not only useful for broad surveillance but also for more targeted evidence gathering. In 2010, a small NGO called SkyTruth used satellite images to document the magnitude of the oil spill caused by the Deepwater Horizon accident in the Gulf of Mexico. The group was the first to publicly challenge British Petroleum's reports of the rate of the spill, showing it was up to five times larger than official estimates.[57] SkyTruth's findings soon made it into mainstream media and congressional hearings and reports.[58] The ensuing Deepwater Horizon criminal case resulted in the largest ever criminal penalty levied against a single US entity as BP pled guilty in 2012 to fourteen felony counts and was fined \$4 billion. SkyTruth has since developed a system of round-the-clock alerts of oil spills using satellite images. The group has also introduced a system providing alerts of illegal bilge dumping, whereby ships bypass costly pollution prevention equipment by simply flushing the bilge water directly into the sea. To detect pairs of vessels meeting at sea, analysts at SkyTruth and Google apply machine learning algorithms to more than 30 billion Automatic Identification System messages from boats to find telltale signs of illicit trans-shipments and correlate these with satellite imagery of oily slicks.[59] 'Once we have reliable data that links pollution to individual ships, we can make information quickly available to actors that have the means to do something about it, either NGOs or government authorities', says the group's founder, Jon Amos.[60]

As the work of SkyTruth suggests, perhaps the most direct way in which digital technologies can enable enforcement of climate regulations is by helping to establish a clear link between specific actions (and actors) and observable environmental harms, thereby facilitating legal action. Beyond satellite, a range of digital tools exist to assist sleuthing (Walsh, 2024). InformaCam is a mobile application developed by WITNESS and the Guardian Project that can be used to turn pictures and videos captured by mobile phones into public proof by using geospatial metadata such as users' GPS coordinates, altitude, compass bearings, light meter readings, and the signatures of nearby devices to verify the exact time, place, and circumstances in which a digital image was captured. Because the project aims at *evidence production* (rather than general monitoring of wider trends or populations) it can be described as a 'data forensic device' (Bitonti, 2024; Glicksman et al. 2016) that helps activists turn dispersed observations into public proof (Eilstrup-Sangiovanni and Sharman, 2022; Suman and Schade, 2021). Satellite-based systems like Climate Action Tracker (described earlier) also play a role in enforcement by enabling climate litigation against specific high-emitting corporations (Setzer and Higham, 2023:19–22). Even where it is not possible to attribute causality in a legally enforceable manner, remote-sensing can serve to identify cases where the probability of non-compliant behavior seems high, guiding further on-the-ground investigations which may usher in legal action.

---

[57] http://skytruth.org/issues/oceans/#sthash.Kxqo8LDO.dpuf.
[58] https://crsreports.congress.gov/product/pdf/R/R41531.   [59] Bladen 2018.
[60] Interview with John Amos, President and Founder of Skytruth, via Skype, 18 September 2020.

A final way technology facilitates enforcement is through widening access to open government data that can help activists hold governments legally responsible for lack of climate action. The last few decades have seen a string of high-profile legal cases in which NGOs have sued governments for failure to heed their commitments under the 1992 UNFCCC or the 2015 Paris Climate Agreement on Climate Change (Bouwer, 2020; Bouwer and Setzer, 2020), or for authorizing environmentally damaging infrastructure (Humby, 2018; Saiger, 2020; Peel and Osofsky, 2018). The first climate litigation case was in Australia in 1994. To date, there have been more than 2,000 cases in the United States alone and 767 outside the United States.[61] The majority of cases have been brought against governments to establish the unlawfulness of administrative policies and actions that fail to take account of national and international climate obligations (Saiger 2020; Setzer and Higham, 2023). An example is the successful challenge by Earthlife of the South African Government's authorization of a new coal-fired power station which rested on the argument that failure to duly consider South Africa's commitment under the UNFCCC and the Paris Agreement made the authorization unlawful.[62]

So far, most climate-change lawsuits have been filed against governments. However, some corporations have also come into the firing line despite not being parties to the UNFCCC or Paris Agreement (Setzer and Higham, 2023). In April 2019, seven Dutch NGOs filed suit against Royal Dutch Shell for failing to align its business model with the goals of the Paris Agreement, despite publicly endorsing the agreement.[63] On May 26, 2021, the Hague District Court granted the NGOs' claim and ordered Royal Dutch Shell to reduce its global CO_2 emissions by 45 percent compared to 2019 levels by 2030 in what is considered the first major climate litigation ruling against a corporation. Shell swiftly appealed the decision, and it is uncertain, at the time of writing, whether the ruling will stand. An optimistic view is that, whether or not the ruling is upheld, Shell has suffered significant reputational damage, and the verdict may serve as a deterrent to other fossil fuel companies.

Yet, despite its growing popularity, climate litigation does not always deliver successful outcomes for NGO claimants. In February 2023, ClientEarth sued the board of directors of Shell plc before the English High Court for failing to take steps to protect Shell against climate-change-related risks. The Court flatly denied the claim. Setzer and Higham (2023:28) find that, as of 2023, only around 55 percent of climate litigation cases had outcomes favourable to climate action. Notably, verdicts favouring NGO plaintiffs are more common in cases of 'climate

---

[61] http://climatecasechart.com (March 30, 2021).
[62] https://climatecasechart.com/non-us-case/4463/.   [63] Kottasová 2019.

washing', which are on the rise in many countries (Setzer and Higham, 2023).[64] Nevertheless, even when courts rule in favour of NGO claimants, defendants often fail to make meaningful amends, questioning the long-term effectiveness of climate litigation (Eilstrup-Sangiovanni and Sharman, 2022). It is also important to acknowledge that litigation strategies, whether to defend indigenous land rights or hold governments accountable for climate pledges, may only deliver change in countries with strong and independent judicial systems.

Risks and Limitations of Technology-Based Monitoring and Enforcement

Climate activists today have greater capacity to collect, analyze, and transmit data on environmental factors than ever before. As we have seen in this section, they use this data in manifold ways to monitor the behavior of governments and corporations and enforce climate policies. Perhaps the most profound way digital technology can empower climate activists is by facilitating 'attribution science'. Scholars of transnational advocacy have long argued that mobilizing people to act against social wrongs requires advocates to present a clear and compelling causal story, with clearly identifiable per-petrators and victims, that can serve to mobilize public outrage (Keck and Sikkink, 1998). A difficulty facing climate activists, however, is that climate change is so complex that it is often difficult to determine who is ultimately at fault. Whilst there are plenty of victims of climate change, identifying those responsible and linking their actions directly to observed harms has often presented an insurmountable challenge. This continues to be the case. However, new digital tools open opportunities for activists to step up moni-toring, investigation, and enforcement actions, thereby reducing information asymmetries and altering power balances between activists, corporations, and states. Today, daily records and images of the entire planet are produced, accessed, shared, and utilized by non-state actors such as NGOs, corporations, scientists, and charitable foundations that operate across borders, reflecting the data's de-concentration and diffusion from governments to other actors (Bennet et al., 2022:735). To borrow James Scott's (2020) phrasing, the ability to 'see like a state' has been extended to non-state actors, ushering in a power shift.

Despite the opportunities afforded by remote sensing and big data analytics for enhancing NGO monitoring and enforcement there are also important risks and challenges. As already discussed, while the costs of commercial remote

---

[64] www.lse.ac.uk/granthaminstitute/wp-content/uploads/2023/06/Global_trends_in_climate_change_litigation_2023_snapshot.pdf.

sensing and computerized data analysis are constantly declining, the financial means and technical expertise required to access these tools still present a barrier for many smaller climate organizations and marginalized communities. For example, many of the indigenous digital mapping projects featured in this section have been initiated by well-financed NGOs and academic geospatial labs based in the Global North, demonstrating the uneven distribution of technological capacities and their correlation with the world's political and economic power centers (Bennet et al., 2022:739). There is danger that as climate activism becomes more tech-heavy it becomes increasingly dominated by well-resourced NGOs and technical experts, crowding out more bottom-up civic engagement and alienating local stakeholders.

Another danger is the growing commercial interest in remote sensing and big data analytics, and the dependence of many digital climate monitoring platforms on funding grants from large tech companies such as Google.org. Despite their non-profit basis, profit motives may lurk in the background of some such platforms, feeding into a culture of 'surveillance capitalism' which thrives on mass collection and commodification of personal data by corporations. Associated with this, there may be personal safety fears arising from the large-scale collection of sensitive data. Extending remote-sensing tools to marginalized communities may encourage them to collect data that can be of strategic value to powerful state and corporate actors, with fatal consequences should the data fall into the wrong hands (Bennett et al., 2022). In one case reported by The Engine Room (2022), data collected by an environmental group about ecological changes in a forested area were misappropriated by hostile actors who used it to violently suppress the local indigenous community.

There is a further worry that (over)reliance on technology may narrow or skew the climate advocacy agenda. Satellite and remote sensing favour a removed, ostensibly 'neutral' perspective over situated knowledge (Bennet et al., 2022:733), which means satellite-based climate activism can risk becoming detached from local needs. At the project level, prior investment in remote sensing capacities may favour a focus on factors that can be documented by satellites and drones (such as changes in canopy cover) over other environmental factors which may be more important to local communities (Bennet et al., 2022). At the organizational level, financial barriers and uneven access to scientific resources currently lead many climate-activist organizations to seek collaborations whereby they team up with universities, professional organizations, or tech companies to conduct monitoring and research. While such collaboration can provide access to state-of-the-art technology, it may also impose constraints on climate organizations' strategic repertoires and limit their ability to set their own agendas and priorities. For example, some activists express concern that 'revolving doors' between climate non-profits and big tech companies mean that civil society actors are less prepared to take commercial actors to task for causing climate harms (Engine Room, 2023).

Crowdsourcing also may not be a silver bullet. The ability to collect and integrate huge amounts of data at relatively low cost has clear benefits in terms of allowing climate activists to quickly learn about, document, and publicize environmental harms. Yet, it also generates new challenges in determining the veracity of vast amounts of data scraped from public websites or crowdsourced from unidentified sources (Muller et al., 2015). Representativeness is another concern, as limited access or financial means to acquire and use digital sensing devices may exclude some individuals and groups. The danger is that climate advocacy based on crowdsourced information or citizen science ends up reflecting the priorities of well-connected individuals with abundant financial means and time on their hands, while being less sensitive to the concerns of wider populations. More broadly, as Morozov (2013:245) notes, 'data-as-truth', or the reliance on algorithmically produced 'truthful' guidelines for decision-making, are based on a scientific tradition that celebrates measurement as seemingly objective. This may obscure persistent inequalities and gaps in what gets measured by whom and what therefore counts as 'truth'. In particular, an emphasis on quantifiable data can lead to poor representation of indigenous cultures and knowledge.

Ultimately, despite its promise to challenge states' information-monopoly, satellite imagery is generally collected and controlled by government agencies or large commercial firms who determine what can be seen, in what format, and by whom, raising doubt about its radical potential (Rothe and Shim, 2018). As Bennett et al. (2022) remind us, despite the seemingly objective nature of satellite imagery, its production and analysis are deeply political: Satellite remote sensing, they emphasize, involves an unequal distribution of resources both on the 'supply side' (those who produce satellite imagery; historically government agencies and now large commercial firms) and on the 'demand side' (those who have the capacity to acquire, use, or train others to use the data for different social and political purposes; often well-financed NGOs and scientific communities in the Global North) (Bennett et al., 2022:730).

Finally, as with digital mobilization and organizing (covered in Section 1), there is a danger that digitalization may reduce public support for environmental monitoring and enforcement, because reliance on technology means that local knowledge and know-how become less and less central to such efforts. As one interviewee reported to the Engine Room (2022:63), 'If you build civic monitoring initiatives around the tech, at some point they will fail. Social interest will drop because you started from the tech and found the problem and applied it to a given community.' A related worry is that growing reliance on digital technologies in climate activism may constrain alternative practices and perspectives. We see a real risk that fascination with the affordances of technology can sometimes lead activists to reject potentially more effective 'low-tech' solutions, or that a focus on trialling new technologies for broad use can lead to local needs being overlooked. The question, then, is not whether satellite and other geospatial technologies are a godsend for climate activism, but rather to what extent, and how, NGOs can adapt to these challenges and harness these technologies for positive change.

Undoubtedly, uneven access to technologies, and dependence on governments, scientific communities, or large tech corporations (even if indirect) will continue to pose challenges for climate organizations seeking to harness remote sensing. Yet, technological infrastructures are constantly evolving. At the time of writing (March 2024), a new satellite, 'MethaneSAT', has recently entered Earth's orbit aboard a SpaceX rocket launched from the Vandenberg Space Force Base in Southern California. MethaneSAT is said to be the world's most advanced methane-detecting satellite and could soon play a key role in combatting climate change by monitoring and publicizing methane emissions from oil and gas fields worldwide in real time and making this data freely available to anyone. Crucially, the $88 million satellite was developed and funded by the Environmental Defense Fund, making it the first satellite to be owned and operated by an environmental non-profit.[65]

Conclusions and Questions for Further Research

The benefits and challenges of employing digital technologies for climate monitoring and enforcement raise important questions for scholars. One question is whether digitalization delivers on its promise of bringing more voices to the table, or merely serves to create new digital divides in the fight against climate change. Another is whether digital monitoring and appeal to 'objective data' get us any closer to building political consensus in a post-truth world, where people are often more likely to accept arguments based on emotions and beliefs than based on facts, and where climate disinformation is ripe. There are further questions about whether a growing emphasis on collecting and analyzing large-scale quantitative data impacts the goals and theories that climate change activists pursue in the sense that 'techno-centricity' means the strategic

---

[65] www.methanesat.org. However, optimism regarding the future ability of climate NGOs to operate their own satellites and thereby achieve autonomous monitoring capacity must be tempered by the observation that launching MethaneSAT required the cooperation of SpaceX and the New Zealand Space Agency, among other corporate and governmental actors.

repertoires of climate groups become increasingly shaped by the ability to collect quantifiable data on target issues.

Finally, scholars might consider whether 'datafication' influences participation in and sustainability of climate activism over time. As we have noted, users of big data are often not climate activists 'in general', but highly skilled data scientists. Compared to traditional activists for whom working on a cause may be a life-long commitment, our interviews suggest that data scientists and software technicians often have 'revolving door' careers where they move between climate non-profits and tech companies or other for-profit enterprises, meaning that their commitment to a given cause may be more fleeting. In turn, there may be a risk that more traditional activists find themselves marginalized and demotivated in an environment that increasingly prices digital skills. These are important questions on which more research is needed to understand how digitalization is impacting climate activism, both in the short and long term, and to enable climate activist groups to benefit from technology.

3 Lobbying

Climate activists are perhaps most visible when they stage mass protests, 'name and shame' corporations, or take direct action to shut down dirty power plants or block new fossil-fuel infrastructure through the courts (see Section 2). So far, our discussion has focused on how climate activists operate 'outside' the political system (and the corporate world) to put pressure on decisionmakers – either indirectly through mobilizing public opinion and organizing communities, or directly through confrontational strategies like direct action and litigation. However, many climate organizations also use so-called 'inside' strategies of direct interaction and dialogue with policymakers (Dellmuth and Tallberg, 2017). Prominent inside tactics include meeting with decisionmakers, offering expert policy advice via policy reports and analyses, and informing decisionmakers about the views and needs of the constituencies climate advocates represent. A wide range of climate organizations have gained access and influence at the United Nations Framework Convention on Climate Change (UNFCCC) and other international policy fora based on their climate expertise (Allan, 2021). Other groups focus more on lobbying decisionmakers at local, regional, and national levels. Their lobbying activities increasingly rely on advanced digital tools.

In this section we briefly explore the impact of technology on strategies of elite lobbying and expert policy advice, before moving on to focus on how digital technologies support organizational formation and fundraising in Section 4.

The importance of political lobbying can hardly be overstated. Influence Map is an independent think-tank that produces data-driven analysis on how big business and finance are impacting the climate crisis. Influence Map maintains a database of 500+ major companies and 250 industry organizations, integrating data on their lobbying activities with emissions metrics to demonstrate how corporate lobbying weakens policies on emissions trading, carbon taxes, clean power, shipping, aviation, and vehicle regulations, and linking these effects of industry lobbying directly to increased GHG emissions.[66] Meanwhile, a growing body of academic research suggests that a major barrier to climate action in many countries is political lobbying by emissions-intensive industries that oppose climate action – particularly oil, gas, and coal (Brulle and Downie, 2022; Downie, 2023). For example, in the United States, more than $2 billion was spent on climate lobbying between 2000 and 2016, with organizations opposed to climate action outspending environmental organizations and renewable energy corporations by a ratio of 10:1 (Brulle, 2018). While better regulation of the political activities of fossil fuel companies and trade associations is crucial to balancing the scoreboard,[67] effective 'counter-lobbying' by the climate movement is equally essential.

We here focus on how digital technologies can serve to reduce information asymmetries and equalize power between non-state climate actors on the one hand, and public policymakers and corporate lobbyists on the other hand. Elite lobbying has traditionally centered on cultivating personal relationships with policymakers through social events like private dinners, celebrity talks, financial contributions to political campaigns or parties, and meetings with Chiefs of Staff and Ministers.[68] Traditional lobbying has also relied on paid advertisements in legacy media and using consultancy firms to prepare detailed policy submissions. These activities all require significant amounts of time, political access, and money. For larger and wealthier interest groups with strong historical ties to policymakers – for example, major industry organizations or trade associations, but also some legacy NGOs – such traditional lobbying methods still constitute an important way of gaining political influence. But for smaller and less established climate organizations on tighter budgets, the digital age has created new opportunities for effective lobbying that relies less on face-to-face contact, established connections, and legacy media ads and more on digitally mediated promotion of political interests directly to individual policymakers.

---

[66] https://influencemap.org.
[67] The 'Global Standard on Responsible Corporate Climate Lobbying' sets out principles for companies and investors to ensure that lobbying efforts are directed towards the attainment of the Paris Goals. https://climate-lobbying.com.
[68] www.adlconsulting.it/en/blog/articoli/the-digital-lobbying-process/.

Speeding Up the Lobbying Process

As for other domains of climate activism, the most direct way digitalization is changing climate lobbying is by rationalizing and speeding up familiar practices through use of information technology and social media for communication. While traditionally, lobbyists might invest large amounts of time and effort in seeking to convince voters to phone a policymakers' office to register their concerns, or seeking to arrange face-to-face meetings with decision makers, today online petitions and flash emails can achieve the same effect cheaper and faster (Vromen et al., 2022; Stürmer et al., 2023). For example, activists can create links that automatically draft email templates for individuals to send to public officials or use social media platforms to organize virtual assemblies that bring together local climate constituencies and decisionmakers. Thanks to such digital lobbying tools, the cost to a lobby group of generating a few hundred email messages from concerned citizens does not differ significantly from the cost of generating thousands. In turn, this means that the reach of an organization's social networks and its ability to mobilize concerned citizens often become more important for gaining political access than organizational budget.[69] On the other hand, the very fact that digitalization makes it easier to generate large volumes of messages from mobilized constituencies may also mean that these messages are more easily discounted by decisionmakers (Cluverius, 2015).

Leveraging Open Government Data

The effects of digital innovation extend beyond reducing the cost of familiar lobbying practices through using social media for communication. The explosion of open government data and advanced data analytics also holds potential to *transform* lobbying processes at a more strategic level (Bitonti, 2024). For example, during the initial phase of a lobbying campaign, digital information-management platforms enable round-the-clock monitoring of policy proposals, legislative changes, social media posts by politicians, public sentiment, and the positions of various stakeholder groups, thereby simplifying the process of monitoring and analyzing political developments. Meanwhile, advanced data analytics enable the use of sophisticated algorithms that can be used to filter, rationalize and fact-check large amounts of information. With many hundreds of policy proposals introduced every year in some jurisdictions, merely compiling information, let alone analyzing draft legislation to understand the possible implications of existing or new proposed regulations, can be a daunting task.

---

[69] https://lobbyit.com/digital-advocacy-in-2024-how-online-platforms-are-reshaping-lobbying/.

Here, web-crawling tools that automatically monitor specific pages of parliamentary, congressional, or other legislative websites and download and systematize information according to project type, sponsors, and stage of progress through the legislative process, can be a real game-changer in helping climate activists stay abreast of political developments – for example, by generating automatic alerts about new legislative proposals submitted on matters of particular interest to them (Bitonti, 2024). In what follows, we briefly consider how digital 'legislative observatories' are changing processes of elite lobbying.

Legislative Observatories

Traditionally lobbying has relied on relatively limited sources of information gleaned from personal connections with policymakers, or 'policy insiders', and from limited surveys and focus groups. In contrast, today's digital lobbyists can draw on much wider information to inform their strategic decisions (Bitonti, 2024; see also Stürmer et al. 2023). 'Legislative observatories' (or also known as 'policy observatories') are digital platforms that use web scraping technology to compile information of interest from the websites of legislative assemblies such as the United States Congress or other national parliaments – automatically and repeatedly.[70] These platforms produce a clear visualization of relevant projects being debated by legislators. Since they eliminate the work of manually downloading and cataloguing information (which can often add up to thousands of pages), they free up more time for legal and impact analysis, or for producing policy reports. Policy observatories can also be used to track the legislative histories of individual politicians, including the proposals they have submitted in the past, their record of voting, and so on. In turn, this can serve to identify opportunities for individually targeted lobbying of policymakers who may be pivotal to a given decision or who may be favourably disposed to act on climate issues.

An example of a legislative observatory is Climate Cabinet Education, an American non-profit that has built the largest database of state and local climate action in the United States, tracking the voting records of more than 5,000 state, city, and county policymakers, and over a million other data points – from clean energy jobs to public health impacts of pollutants.[71] Climate Cabinet Education integrates this data to produce a 'climate heatmap' that identifies critical leverage points for influencing political decisions on clean energy and environmental justice. By tracking records of support for

---

[70] www.ilo.org/wcmsp5/groups/public/–ed_dialogue/–act_emp/documents/publication/wcms_887579.pdf.

[71] https://climatecabineteducation.org/.

green policies and providing contact information for local and state decision makers, the climate heatmap provides journalists, activists, and the wider public with crucial information about what climate legislation is being considered in state legislatures across the United States, and how to identify political leaders who are key to influencing these decisions. As Climate Cabinet's website states, 'State and local leaders are critical for solving the climate crisis. Yet with 500,000 local policymakers throughout America, it's hard to know where to start to make the biggest climate impact now. At Climate Cabinet, we make this numbers game easy ... We've built the largest database of state and local climate action ... to find leaders with outsized opportunities to act on climate and environmental justice. Then, we activate our team of clean energy professionals, policy experts, former caucus staff, data scientists and organizers to make sure these key leaders have the policy support they need.'[72] The website highlights efficiency gains from using digital technologies to identify opportunities for targeted lobbying and policy support: 'Fossil fuel interests are rich and vastly overpower clean industries. We can use automation to do more with less. We're connecting the dots and building shared resources for the movement.'[73]

Another climate-focused policy observatory, with more global reach, is Climate Policy Radar.[74] This initiative, which is led by a team of international experts in climate-change law and machine learning and supported by several British universities and scientific institutions, uses advanced data science and AI 'to map and analyze the global climate law and policy landscape to provide a comprehensive overview of what climate change interventions are being implemented, by whom, and where, and to model the potential impact of different climate laws and policies.' Climate Policy Radar's database consists of national submissions to the UNFCCC along with thousands of laws, policies, strategies, and action plans from national and subnational governments in different countries, providing a powerful source of information for climate activists around the world to assess political proposals, highlight insufficient government action, and lobby for new policies. Yet another relevant platform is the policy observatory of the Open Future Foundation, which tracks European legislation with potential to advance openness and contribute to a digital public space maximizing the social benefits of shared data, knowledge and culture.[75]

---

[72] https://climatecabineteducation.org/about-whoweare/.
[73] https://climatecabineteducation.org/about-whoweare/.
[74] https://climatepolicyradar.org/what-we-do.   [75] openfuture.edu/observatory.

Engaging Policymakers with Digital Tools

Moving on from policy monitoring and analysis to the design and implementation phase of lobbying campaigns, digital analytics can aid in strategic prioritization by identifying specific opportunities for lobbying decisionmakers and forecasting expected results (Bitonti, 2024). As Bitonti explains, much like A/B testing can help activists mobilize support for mass protests or other campaign actions (see Section 1), AI algorithms can run simulations and evaluate the likely outcomes of different lobbying campaigns and tactics, using publicly available data such as the outcomes of earlier campaigns, and previously recorded votes by members of a legislative body. Digital monitoring and data analysis can also assist advocates in building strong partnerships and coalitions by identifying like-minded stakeholders (Bitonti, 2024).

Shaping political agendas not only requires climate advocates to pinpoint crucial decision-points, and frame issues in ways that can break through political barriers and capture decisionmakers' attention. It also requires them to present credible evidence to make a compelling case for political change. Here, new information sources like satellite and remote sensing data, open government data, and citizen science projects (described in Section 2) have made it easier for climate lobbyists to evaluate the impact of alternative policies and to present a case for new climate laws and regulations, based on strong evidence. This has, in turn, helped to narrow the gap in expertise and knowledge between non-state climate advocates and public policymakers or state-backed scientific institutions. More broadly, as Fung and O'Rourke (2020) note, publicly available 'big data' can help to level the playing field in pitched interest-group battles over environmental issues. To fix ideas, open government databases such as the E-PRTR (see Section 2), and online databases of national submissions to the UNFCCC give activists direct access to government-legitimated information about political pledges, targets, and regulatory initiatives (and their effects on target actors), which activists can use and repackage to bolster their proposals for new legislative and regulatory measures (Fung and O'Rourke, 2020).

Finally, digital tools such as infographics and 3D maps can enhance lobbying campaigns by helping advocates to present complex data in easily understandable formats. Effective political lobbying often depends on making complex problems intelligible to busy decisionmakers and presenting clear, actionable solutions. Here, advanced data analytics can help lobbyists transform complex data into more accessible and user-friendly information that allows policymakers to grasp complex problems and proposed solutions more easily.

Of course, the same digital tools that benefit climate advocates also benefit large business organizations or fossil-fuel lobbyists. Still, in a context where corporate actors and business associations typically have greater financial power and enjoy more direct political access than most NGOs and civil society groups, digital tools may help to narrow the gap in access to information, expertise, and analytic capacity, thereby opening the door to more effective lobbying by a wider range of progressive climate change actors.

Limitations and Risks of Digital Lobbying

The potential limitations and dangers of relying on digital technology to support lobbying are similar to those identified for general digital monitoring and enforcement activities: Using AI and big data analytics to identify urgent political issues and gauge potential support for different legislative solutions may mean that local concerns, or issues at the margin of mainstream political debates, recede into the background. Policymakers may tire of being bombarded with 'boilerplate' email and social media messages from concerned citizens and their representatives, and the lack of 'authenticity' of such messages may make it harder to judge the real extent and depth of public concern. Possibly, zoom-meetings and digital assemblies cannot meaningfully replace personal ties to policymakers forged at private dinners or other face-to-face social events. Individual climate lobbyists may also feel alienated by a growing emphasis on computerized work-processes and digital skills. On the other hand, compared to climate activists at large, many elite lobbyists, including those working for non-profit organizations, may see lobbying more as a career choice and therefore be more willing to invest in acquiring the necessary digital skills to stay at the forefront of developments in their profession.

It would be amiss not to acknowledge that corporate lobbying by carbon-intensive manufacturing industries, chemical industries and fossil fuel companies vastly outstrip the lobbying activities of pro-climate groups (InfluenceMap, 2024). These industries have a long history of seeking to undermine climate-friendly regulations and exaggerating the potentially negative economic consequences of green policies. Like many climate-activist groups, these companies are adept at exploiting digital lobbying tools, but also have large numbers of highly paid and professionally trained lobbyists permanently stationed in places like Washington, Brussels, and other centers of global policymaking. It thus continues to be the case that big money buys wider access and greater potential influence on policymaking. Thus, far from being an *even* playing field, elite lobbying may perhaps be better described as a field where climate activists need to think pro-actively and creatively about how to harness digital tools to play at all.

Conclusions and Questions for Further Research

Very little research has examined the effects of using legislative observatories or other digital platforms to design and manage lobbying campaigns, and we are not aware of any studies that explore how digitalization is changing the landscape of *climate* lobbying, specifically. Indeed, much research on lobbying strategies still tends to focus on a largely outmoded set of pressure strategies, ranging from letter-writing campaigns to phone calls, or face-to-face meetings with decision-makers (Chalmers and Shotton, 2016). Hence, this is an area where more research is urgently needed. Specifically, we see a need for comparative work on how digital lobbying works in different political systems and cultures. Scholars should look at how climate lobbyists operate in some of the biggest democracies (India and Indonesia) as well as the largest and most entrenched authoritarian regimes (Russia and Iran), which tend to be understudied when compared with democracies in the Global North. There is also a need to reflect on whether and how the profiles and career pathways of climate lobbyists are changing due to new technologies.

4 Forming, Fundraising, and Networking

In 2013 a conservative Australian government, led by Tony Abbott, decided to abolish the government's Climate Commission. The response was fast, and furious, as climate advocates decided to fundraise for a new and independent Climate Council. Rather than knocking on doors or launching a TV advertisement campaign, they ran a crowdfunding campaign. In just over a week, in the biggest crowdfunding campaign in Australian history at the time (McLean and Fuller, 2016), they fundraised $1.1 million from 16,000 'founding friends' (Climate Council, 2023). The Climate Council was established as a not-for-profit charity with a mission 'to be a courageous catalyst propelling Australia towards bold, effective action to have emissions plummet during the 2020s and hit net zero by 2035'. As this example illustrates, digital technologies can enable new organizations to be launched at low cost, and very quickly.

This section explores how the internet and social media platforms have made it easier to establish and fund climate activist organizations with both local and global scope. While International Relations literature has been relatively silent about how NGOs form and fundraise, there is a large scholarship in political science on the importance of digital technology in funding party-political campaigns (Gerbaudo, 2018; Gibson et al., 2014; Kreiss, 2016; Karpf, 2012). Meanwhile, scholars of social movements and transnational networks have focused on how social media enables 'leaderless' social movements to form, but not on the implications of these platforms for formal organizations (Gerbaudo, 2012; Shirky, 2008). Yet, to understand the organizational ecology of contemporary climate advocacy, it is crucial to understand why particular organizational forms emerge, and when; why first movers may have an advantage, over latecomers; and which organizational forms thrive in different environments.

This section examines how digital technologies have changed the process of founding new organizations, raising funds, and networking among organizations. As in other sections, our focus is specifically on climate organizations, although other advocacy organizations are also likely to benefit from the technology described here. As explained in the Introduction to this Element, we believe that the dynamics of organizational formation, funding, and networking in the climate advocacy sector are particularly important to analyse since we have seen the emergence of a wealth of new climate activist groups in recent years, making use of digital tools.

Organizational Formation: Recruiting Members and Establishing an Organization

Transnational advocacy scholars have had relatively little to say about the costs, risks, and processes of setting up new advocacy organizations, but have instead tended to focus on the strategies, tactics, and networking patterns among advocacy organizations once established. This is surprising, given the large number of new climate advocacy organizations emerging in the last decade – from 350.org to Extinction Rebellion, Fridays For Future, and Last Generation. Digital technology facilitates the establishment of new organizations and movements – even if it is not the decisive factor in why new organizations emerge. Social media enable activists to find each other more easily and coordinate online (Shirky, 2008; Margetts et al., 2015). One tweet or Facebook post can reach thousands of people and encourage others to take action (Margetts et al., 2015; Dennis, 2019), as illustrated by Greta Thunberg's lone protests outside the Swedish parliament which inspired thousands to replicate her actions. Budding climate activists can connect via social media and cooperate to develop new organizations like Fridays For Future. Although not all social movements become formalized advocacy organizations (some remain diffuse, loose networks), digital technologies have enabled social movements to scale more quickly (Tufekci, 2017; Hall, 2022). Founders of new climate organizations frequently use social media, email communications, and online petitions to recruit new members. Digital advocacy organizations like 350.org, MoveOn, Campact, and GetUp! also pioneered new forms of membership (Karpf, 2012). Rather than requiring members to pay annual subscriptions or attend regular meetings, they simply ask members to sign up to email communications, and encourage them to sign online petitions and share social media posts. This lowers barriers to setting up an advocacy organization, although it also means that members may not identify as closely with the organization (Karpf, 2012; Fisher, 2019; Hall, 2022).

Technology has also reduced operational costs for climate advocacy organizations (Karpf, 2012). The internet lowers transaction costs of recruiting volunteers and coordinating among staff. Organizations can run low-cost events online whether podcasts, live-streamed discussions, or zoom calls. Advocacy organizations no longer need to pay for fixed office space but can ask staff to work from home on their own computers. Some organizations remain fully remote throughout their existence; for example, the Engine Room is a 'global, fully-remote team'[76] as is MoveOn (Karpf, 2012). However, many activists we interviewed also highlighted the importance of in-person meetings with colleagues, volunteers, and supporters (Hall, 2022).

Fundraising

'It's hard to remember, but in 2008 it was very difficult to accept donations globally.'[77]

–Nathan Woodhull (Software Engineer, Community Organizer, and Founder of *Control Shift Labs* and *Daisy Chain*)

Online donations are one of the most popular ways for individuals to fund NGOs today. Yet, as Nathan Woodhull noted, less than twenty years ago it was difficult for advocacy organizations to fundraise from multiple countries online as it required the creation of custom software.[78] For example, Wikipedia had to write their own software to fundraise internationally from members.[79] By contrast, today many current climate organizations rely predominantly on online donations from members. It is cheaper to seek donations via email or social media than hiring street canvassers, sending out letters, or making televised appeals. Some organizations also use digital analytics to test the subject, content, formatting, and timing of fundraising emails to maximise returns. For example, MoveOn has tested which colour 'donate button' earns most donations (Karpf, 2016). MoveOn also calculates how much money an email is likely to raise if it is sent to everyone at once, versus to a smaller segment of the list at a time (Hall, 2022). In future, AI may be useful in generating more effective, and cheaper, fundraising appeals. For example, the Good Data Institute has developed an app, 'Chuffed GPT', which uses Chat GPT to create targeted campaign content and emails to supporters. Staff simply enter a prompt, and the app produces a complete draft of a fundraising email.[80]

---

[76] The Engine Room 'Jobs' www.theengineroom.org/jobs/.
[77] Interview with Nathan Woodhull, 22 November 2023.
[78] Interview with Nathan Woodhull, 22 November 2023.
[79] Interview with Nathan Woodhull, 22 November 2023.
[80] Good Data Institute 'Chuffed GPT: An Algorithm for Altruism', www.gooddatainstitute.com/post/chuffed-gpt-an-algorithm-for-altruism (last viewed 13 December 2023).

Across the world there is a growing tendency towards donating to organizations online. Over fifty percent of individual donors prefer to give online with a credit or debit card over other forms of donation (direct mail/cash/bank or wire transfer or PayPal) (Nonprofit Tech for Good, 2018:6). There is, however, significant global variation in how people prefer to give. Africa has the highest rates of donors who prefer to give via mobile money (9 percent), as mobile apps to transfer funds between individuals have been around for a long time (e.g., Mpesa in Kenya). In Asia, 52 percent of donors prefer to give online with a credit or debit card; 13 percent with cash; and 2 percent via mobile money.

Another important trend noted in the 2020 Global Giving Report is that 49 percent of individuals give recurring donations – whether weekly, monthly, quarterly, or annually (Nonprofit Tech for Good, 2020). These regular donations give organizations more financial security and enable longer-term planning. The Global Giving Report data does not focus specifically on climate advocacy organizations, nor does it include other forms of financing (say, government, foundations, or corporate donations). Nevertheless, these trends in individual donations are important, given that many new environmental organizations do not accept donations from corporations or governments. What's more, getting a large pool of supporters to give small amounts of money may increase public support and legitimacy as more people feel directly involved, and as people can directly 'buy into' specific projects.

In sum, digital platforms have widened the funding models available to climate organizations. This may be particularly important for organizations operating in hostile climates such as repressive countries where NGOs are barred from accepting foreign funds (Chaudhry, 2022) or where there is high density of organizations, resulting in stiff competition for funding from foundations and governments (Bush and Hadden, 2019).

There is a need for further research on the financing of climate advocacy organizations and on changes to funding patterns over time, particularly as there are large differences in organizational budgets (see Figure 5). Interestingly, the organizations that have the least funding (for example, Extinction Rebellion or Sunrise Movement) are often seen as the most effective at mobilizing and gaining media attention, compared to older, more established NGOs like Greenpeace or WWF (Ozden, 2022). Sunrise Movement relies mostly on volunteers and focuses almost exclusively on advocacy, while WWF has paid professional staff and dedicates much of its resources to delivering conservation projects around the world, rather than advocacy.[81] This is not surprising: NGOs tend to professionalize over time, and as a result require more funding to support bureaucratic functions such as grant writing, reporting, and compliance).

---

[81] WWF-US. 2023. *Annual Report*, https://files.worldwildlife.org/wwfcmsprod/files/FinancialReport/file/8yisj3spgl_WWF_AR2023_12_20.pdf.

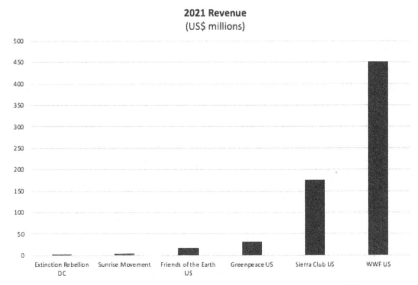

2021 Revenue
(US$ millions)

Figure 5 Revenue of US environmental organizations.[82]

Coordinating and Networking

Digital technology has been pivotal in connecting climate activists both domestically and internationally. Scholarship on transnational advocacy in the 1990s highlighted how fax, email, and cheaper flights enabled denser transnational networks to form (Keck and Sikkink, 1998). Since then, technology to support international networking has evolved dramatically. When 350.org was formed in 2008 they did all their coordinating via email and skype, which was a giant leap forward from connecting via letter, fax, or phone in terms of increasing speed and reducing costs. However, as Jon Warnow, the first lead digital campaigner for 350.org explained to us: 'we struggled just to do basic stuff on Skype in the aughts in 2008, 2009, and honestly, it wasn't until . . . maybe like 2015, 2016, when this stuff [online video conference calls] actually became much more usable.'[83] Warnow explained that Zoom enabled a breakthrough in building trust within 350.org's networks as 'there's nothing like being able to see people's faces to enable really good collaboration across borders, across cultures, all that stuff.' Since the COVID-19 pandemic, numerous other digital tools and platforms, such as Slack, have become common tools for coordination and networking within and between climate advocacy organizations.

---

[82] This graph is produced by the authors, based on each organization's Annual Report and/or IRS 990 Form.

[83] Interview with Jon Warnow, 22 November 2023.

Technology has also facilitated the growth of large inter-organizational networks, from the Climate Action Network to 350.org. An illustrative example is the Fossil Fuel Non-Proliferation Treaty (FFNPT), which was established in late 2019. The FFNPT had to operate largely online for the next two years due to the Covid pandemic, but thanks to digital technology, especially Zoom and Slack, participants forged a large and diverse network including 80 cities and subnational governments, over 5,000 individuals and hundreds of civil society organizations across the world.[84] The network has members in almost every country in the world, and has also built alliances with indigenous peoples in the Amazon and Pacific Island leaders. Additionally, FFNPT activists have coordinated open letters from 101 Nobel laureates; 200 health institutions, 3,000 scientists and academics; and faith institutions representing more than 1.5 billion people.[85] They coordinated many of these initiatives during the pandemic, when travel in person was extremely difficult, relying largely on digital technology to grow the network.

Challenges and Risks of Online Founding, Funding, and Networking

Despite the many benefits, there are also risks and limitations entailed in building coalitions and networking online. First, social media often cannot replace the power of meeting in person. Research has found that activists build more trust and stronger alliances when interacting face to face compared with online (Hall, 2022). Second, as already discussed, not everyone has access to cheap, reliable internet (Schradie, 2018; 2019). Hence, some individuals and groups are more likely to participate in online networks than others. Third, as we discuss in Section 5, there may be risks to forming transnational alliances online, as governments can collect data on online activities and repress or restrict activism (Chaudhry, 2022; Chaudhry and Heiss, 2022; Dauvergne, 2020; Feldstein, 2021; Gohdes 2020). There are many examples of environmental activists being targeted, even murdered, by state authorities that use facial recognition cameras and machine learning to suppress dissent. In 2017, in the Philippines alone, there were 48 documented murders of environmental and land activists in which state authorities were likely involved (Dauvergne, 2020:162). Finally, just as it benefits pro-climate activists, digital technology for coordinating and

---

[84] Fossil fuel treaty, 'Endorsements' https://fossilfueltreaty.org/endorsements (last accessed 24 April 2021).

[85] Far-right politicians like former President Donald Trump, and former Brazilian President Jair Bolsonaro and their followers, have used Twitter to attack activists like Thunberg online.

networking can, obviously, also be harnessed by groups seeking to obstruct climate action. Groups that spread conspiracy theories, undermine climate science, and question policies to tackle climate change are also using technology to fundraise, recruit members, and network transnationally, and are harnessing social media platforms like X (previously Twitter) to attack climate activists online.[86]

Conclusions and Questions for Further Research

Digital technologies have enabled new climate advocacy organizations to form, fundraise, and network at a wider scale than ever before. However, we need further research on population-wide impacts of digital technology. Have more climate activist organizations been founded in the digital era thanks to the ease of finding members and potential donors? Has there, for example, been an increase in crowd-funded organizations? Or are other factors driving population growth, for example, a younger generation's frustration with established environmental NGOs and their lack of urgency and radicalism in tackling climate change? This would be consistent with the trend of new groups – like XR, FFF and Last Generation – pioneering more radical actions. Is the growth of organized climate activism a lasting trend or a 'fad' triggered by temporary enthusiasm for the possibilities offered by new technologies? We know that the creation of new international NGOs focused on environmental conservation has stagnated in the United States (Bush and Hadden, 2019) but is the same true for *climate* organizations in the United States and elsewhere?

We also do not know enough about how newer climate activist organizations are funded, and the extent to which funding models differ between newer and older generation groups. This is an important area for further research given that funding not only assures organizational survival, but also determines who organizations are reliant on and accountable to.

Another set of questions pertain to the implications of the rise in crowdfunding for organizational strategies, legitimacy, and impact. If climate organizations are less reliant on large philanthropic donors, does this allow them to adopt more radical goals, strategies, and tactics? Are the newer generation of climate advocacy organizations, which have largely recruited their members online, more effective at mobilizing the public than an older generation of professionalised environmental NGOs? Or do members who have signed up online have a

---

[86] Right-wing groups and individuals are much more likely to be climate-sceptic than those on the left (Dobson, 2016:92; Lowles, 2021:57). Although some far-right activists have also cited climate action as part of their rationale for atrocious attacks against immigrants; and justifying violent terrorism (Lowles, 2021).

lower sense of affiliation with these organizations and therefore less willingness to take action? Examining how advocacy organizations recruit staff, volunteers, and members and how they fundraise is important for understanding these organizations' robustness and potential impact over time.

5 The Dark Side of Digital Technologies

While it can support climate activism, modern information and communication technology can also be part of the problem. Digital technologies have a significant carbon footprint, climate activists face growing digital surveillance and online attacks, and social media platforms are found to spread disinformation about climate change. Meanwhile, many communities at risk from climate change lack access to basic digital resources, from reliable internet access to trustworthy online information in their own language (Engine Room, 2023). These downsides mean that, for some, initial optimism about digitally empowered activism as a tool to build broader social movements and bypass state control has given way to pessimism. Brett Solomon, founder and executive director of Access Now, a global non-profit digital rights organization, puts it bluntly: 'the internet has stopped being our friend and is increasingly becoming our enemy. The balance has shifted. Tech often weakens activists' ability to achieve their goals'.[87]

Digital Repression

Just as the advent of digital media has broadened opportunities for political activism, it has also increased the instruments available to states to repress their citizens by surveilling, manipulating, and censoring digital information flows (Gohdes, 2020; 2024; Crawford, 2021). Common repression tactics include blocking platforms, disabling VPNs, forcing the release of user data from online platforms, or relying on biometric surveillance such as facial scans to identify and track dissenters (Funk et al., 2023; Earl et al., 2022). Some states regularly shut down the internet to quell public dissent or create their own fenced-off versions that they tightly control (Gunitsky, 2015). For example, China's 'Great Firewall' – a combination of legislative measures and technologies used to regulate the internet – aims to keep foreign content out, and the Chinese government reportedly spends $6.6 billion per year monitoring what goes on inside (Crawford, 2021).

A key concern is that digitalization allows for more targeted violence against activists. Recent studies find evidence of growing targeting of individual activists via digital platforms where they're first identified, subjected to trolling, in some

---

[87] Interview with Brett Solomon, 28 November 2023.

cases doxed,[88] and in extreme cases, tracked down and killed (Engine Room, 2023; Earl et al., 2022; Gohdes, 2024). A report by Freedom House (2023), 'Freedom on the Net', found that in fifty-five of seventy countries surveyed, people faced legal repercussions for expressing themselves online, while there were examples of people being physically assaulted or killed for their online commentary in forty-one countries. Some of the groups most vulnerable to online repression and state violence include indigenous climate activists seeking to protect their lands against logging and mining, and small island state activists campaigning against increased emissions.[89] As COP28 kicked off in Dubai in 2023, many climate groups were warning that the United Arab Emirates Government would target activists with spyware.[90] However, digital repression is not limited to autocracies. For example, British police have monitored social media to make preemptive interventions and arrests before protests occur, while US police have used Stingray surveillance devices to monitor protesters' mobile phones (Earl et al., 2022:2).

While state repression in the digital era shares many similarities with predigital repression – most obviously the fundamental aim of limiting and controlling information – Earl et al. (2022) highlight how digital technologies can change both the aims and scope of information control. Whereas predigital censorship focused on suppressing knowledge and changing beliefs by preventing undesirable information from circulating, digital censorship further suppresses individuals' ability to communicate with and be seen by others (Earl et al., 2022). This acts as a barrier to community building, social mobilization, and campaigning. As Brett Solomon of 'Access Now' reflects, 'the targeted use of spyware, take Pegasus as an example ... it's like weed-killer. It kills speech, privacy, association, it kills everything'.[91] Notably, Access Now was created with the specific aim of helping activists who face online intimidation and attack. The group operates a 24/7 'digital security helpline' for activists in nine languages, which states: 'If you're at risk, we can help you improve your digital security practices to keep out of harm's way. If you're already under attack, we provide rapid-response emergency assistance'.[92] This is one way activists can adapt to and circumnavigate repression (Gohdes, 2024).

---

[88] Doxing involves publicising personally identifiable information about an individual or organization, usually via the Internet, without their consent.
[89] Interview with Brett Solomon, 28 November 2023.
[90] www.accessnow.org/press-release/letter-governments-cop28/.
[91] Interview with Bret Solomon, 28 November 2023.
[92] AccessNow, 'Helpline', www.accessnow.org/help/ (accessed 21 November 2023).

Disinformation

Fear is not the only mechanism of digital repression. Online censorship can also impose costs on activists by making information more difficult to access, or by spreading false information to confuse users about what is true and false. Climate-related disinformation continues to spread online through social media platforms that fail to moderate content. For example, new technical tools such as automated bots and AI-powered algorithm generators have been found to shape climate discussion on major online platforms like Twitter/X (Earl et al., 2022). On social media, emotive content, posts shared by friends, and algorithmic recommendations can also help false information gain traction. A recent report by Climate Action Against Disinformation found major fossil-fuel companies were behind much misleading information online (CAAD, 2023). The report revealed that fossil-fuel companies paid Meta, which owns Facebook, Instagram, and WhatsApp, for climate disinformation advertisements to the tune of $6 million per year.[93]

Another source of online disinformation is organized climate denial. A 2020 report by Influence Map found that Facebook earned an annual revenue of $68 million from disinformation ads posted by known climate denier groups.[94] Prominent examples include 'Watts up with that', 'Global Warming Policy Forum', 'Climate Realists', and 'Friends of Science' (Drieschova, 2023). So far, scholars have found climate sceptics to be less successful at mobilizing large followings online than pro-climate activists (Drieschova, 2023:241). But this could change. Generative AI threatens to supercharge online disinformation, as the use of AI-generated images, audio, and text make it easier, faster, and cheaper for mis-informers to distort the truth (Funk et al., 2023). Importantly, false information tends to rise on social media sites around key political events such as COP climate talks, or in the wake of extreme weather events. For example, during the European 2023 heatwave, posts which claimed that arson [by migrants] was responsible for the wildfires gained thousands of retweets (CAAD, 2023). Other social media posts alleged the manipulation of weather maps by TV broadcasters. The negative impact of such disinformation should not be underestimated. As one activist reflects, 'We can spend 10 years building trust, that can be dissolved in 10 minutes by a fake video'.[95] According to the UN Intergovernmental Panel on Climate Change (IPCC), disinformation represents a significant barrier to climate action by undermining accurate

---

[93] Estimates from 1 January – 31 October 2023 (CAAD, 2023, p. 43).
[94] https://influencemap.org/report/Climate-Change-and-Digital-Advertising-86222daed29c6
f49ab2da76 b0df15f76#1.
[95] Interview with Brett Solmon, 28 November 2023.

transmission of climate science and by thwarting open debate necessary to build support for ambitious climate action (IPCC, 2022:56, 1577).

Along with spreading false information, digital media also enables other forms of repression such as 'flooding' and 'channelling' whereby repressors make paying attention to certain information more attractive (Earl et al., 2022:7–8). Flooding is a method of manipulating information and opinion by introducing massive amounts of information into the digital space. Authoritarian governments have been found to mass-produce online content using automated bot accounts which fabricate social media content as if it were the views of ordinary people to drown out criticism (ibid.). Online information can also be used to distract. Ahead of COP28 in Dubai, the United Arab Emirates released a game in Google App Store: 'Immerse yourself in the urgent world of global climate action in Climate Guardians COP28. As a delegate at the UAE conference, craft policies, solve environmental challenges, and collaborate with players worldwide. Explore stunning UAE locations, make impactful decisions, and compete for the highest cooperation score. Join the movement for a sustainable future today.' Critics were quick to suggest this game would be used not only to distract but also to spy on users.

Is climate disinformation getting worse? Seemingly, yes. #ClimateScam, a hashtag frequently used to promote denialist and conspiratorial content, has become more prominent on X/Twitter since Elon Musk took over the platform in 2022 (CAAD, 2023). A study by Global Witness which polled climate scientists worldwide found that 39 percent had experienced online harassment (CAAD, 2023). Many social media platforms have policies in place to counter misinformation or online abuse. For example, Google announced in 2021 that it would block publishers found to be spreading climate change disinformation from accessing its advertisement products.[96] YouTube, Meta, and TikTok have also vowed to address climate misinformation (CAAD, 2023). However, enforcement is found to be lacking.[97] Analysis by the UK-based Centre for Countering Digital Hate said Facebook did not add fact-checking labels to half of the posts pushing content from prominent climate change deniers (Center for Countering Digital Hate, 2022).[98] In 2022, a coalition of more than 450 scientists called on the executives of major advertising and public relations firms to drop their fossil-fuel clients and stop what the scientists said was their continued spread of disinformation around climate change.[99]

---

[96] https://support.google.com/google-ads/answer/11221321?hl=en.
[97] www.nytimes.com/2023/05/02/technology/google-youtube-disinformation-climate-change.html.
[98] https://counterhate.com/blog/facebook-failing-to-flag-harmful-climate-misinformation-new-research-finds/.
[99] www.reuters.com/business/cop/scientists-target-pr-ad-firms-they-accuse-spreading-disinforma tion-2022-01-19/.

Digital Divides

Yet another concern regarding digital climate advocacy is that it leaves behind large segments of society. The digital revolution is widely seen to empower individuals by enabling them to access and share information beyond their local communities or national borders. However, the uneven distribution of digital tools and competencies risks creating new digital divides, widening the gap between cosmopolitan, wired urban elites and wider populations who may lack such skills (Lynch, 2011:307). Unless it is provided in a format that is easy to understand and navigate for the non-expert, open government data and other open-access databases can risk magnifying digital divides by empowering elites with expertise to navigate such systems, while remaining unaccessible to non-experts.

In terms of how digital divides intersect with other societal fault lines, research has found that global internet use is lower for women than men, especially in low-income countries (Lythreatis et al., 2022; World Bank, 2023), and that rates of internet access and 'digital literacy' tend to be higher in richer countries, in urban settings, and among socio-economically well-off groups. As of January 2024, there are 5.35 billion internet users worldwide, accounting for 67.1 percent of the global population (DataReportal, 2024). Against this global backdrop, digital progress has been uneven, exacerbating the gap between the digital haves and have-nots. According to World Bank and ITU data, 90 percent of the population in high-income countries are online, compared with just 44 percent in developing countries. Globally, 72 percent of households in urban areas has access to the internet, almost twice as many as in rural areas (38 percent).[100] Broadband in wealthier countries is five to ten times faster than in low-income countries.[101] Effectively, this means the populations most at risk from the impacts of climate change often have the least access to information about it and also less access to digital early-warning systems for climate emergencies.[102]

Extractivism

A final 'dark side' of digitalization is that it contributes directly to climate change by depleting natural resources and producing GHG emissions (Crawford, 2021). Colour displays, speakers, camera lenses, rechargeable batteries, hard drives, fibre optics, and other key elements in digital

---

[100] www.itu.int/net4/wsis/forum/2022/Agenda/Session/468.
[101] https://blogs.worldbank.org/en/voices/digital-era-all.
[102] www.adaptation-undp.org/bridging-digital-divide-will-save-our-planet.

communication systems all rely on rare earth minerals. Lithium mines in Nevada, southwest Bolivia, The Congo, Mongolia, Indonesia, and Western Australia all testify to the rampant 'extractivism' feeding the digital economy (ibid.). Once produced, digital communication systems are often powered by fossil fuels. For example, the large data centres where many large language models are trained and deployed are extremely energy-intensive, collectively accounting for about 1 to 2 percent of global electricity usage (Li et al., 2023; Crawford, 2021:43). Some forecasts suggest that by 2030, the electricity usage of communication technologies could account for nearly a quarter of global GHG emissions (Andrae and Edler, 2015). At the end of the commodity chain is toxic waste: with over six billion new ICT goods sold annually, e-waste has become the largest waste stream in many countries, with developing countries bearing a disproportionate burden as e-waste is illegally shipped there from other countries (Global Information Society Watch, 2020). In sum, while digital technologies can help to promote climate-friendly products and policies, these technologies themselves are often far from climate-friendly.

Organizational Adaptation?

So is digital technology a boon or a curse for climate advocacy? Our aim in this Element is not to assess whether digital technology helps or hinders pro-climate activism overall, but to consider how climate advocacy organizations adapt to new technologies. In this context we note that while they are subjected to growing surveillance and online threats, activists continue to develop more sophisticated strategies to protect themselves and circumvent information controls. Digital rights groups have emerged which promote digital security for activists through capacity-building and awareness-raising and by developing best practices for safe communication and data storage (Engine Room, 2023). Digital rights organisations such as Access Now, European Digital Rights, and Electronic Frontier Foundation have also opened digital security helplines which offer emergency assistance to individuals facing online harassment, attacks, and censorship. Still other groups, such as the Goethe Institute in sub-Saharan Africa, specialize in enhancing digital literacy. In sum, a new organizational ecology is taking shape in response to digital repression.

Digital pessimists cite digital repression to argue that the internet represents a net negative for social movements. However, there is little scholarly consensus on the impacts of online repression (Earl et al., 2022:8). Hobbs and Roberts (2018) find that when governments impose censorship on previously uncensored information, citizens are incentivized to learn about or develop new methods of censorship evasion. For example, Chinese social media users reacted to the government's block of Instagram by acquiring private virtual networks, and users were subsequently more likely to join censored websites like Twitter (X) and Facebook (Hobbs and Roberts, 2018). Similarly, Pan and Siegel (2020) find that while digital repression may suppress dissent by individuals who are directly targeted, it does not tend to deter their online supporters, who instead become more ready to engage in online dissent. Still other studies have found that awareness of censorship, especially during crises, pushes users to find ways to circumvent it and incentivises them to seek out concealed information (Lynch, 2021). Conversely, when users are unaware censorship exists, compensating for information manipulation becomes very difficult, especially in online contexts where censorship is masked by algorithms (Lynch, 2021).

Besides hitting back against digital repression and censorship, climate activists also seek to offset the negative environmental impacts of the digital technologies they depend on. The Green Web Foundation, a Dutch NGO, provides an open-access database that measures the carbon emissions of websites and cloud services and helps organizations switch their hosting to green providers.[103] In the last decade, the 'right to repair movement' has pushed for the transition to circular consumption of electronics, based on 'repair and reuse' principles. As an example, the non-profit foundation, Repair Café, which started in the Netherlands in 2007 builds skills to repair digital devices. As of 2024, there are more than 3,000 Repair Cafés in more than 25 countries worldwide.[104] Other examples include Right to Repair Europe,[105] a coalition representing over 100 organisations from twenty-one European countries, The Restart Project based in London,[106] and the Association for Progressive Communication which helps climate organizations to connect in more sustainable ways.[107]

Activists are also trying to address extractivism and surveillance capitalism through mobilizing dissent. Tech Workers Coalition organized a walk-out in September 2019, which saw millions of tech workers join the youth #climatestrike under the banner #TechClimateStrike. The coalition seeks to highlight how the tech industry cultivates a 'green' public

---

[103] www.thegreenwebfoundation.org.   [104] www.repaircafe.org/en/visit/.
[105] https://repair.eu/.   [106] https://therestartproject.org/.
[107] www.apc.org/en/pubs/guide-circular-economy-digital-devices.

image but is in fact a major contributor to climate change and profits from selling surveillance technologies. Another example is Climate Action Tech, a global community of tech workers organized on Slack which seeks to seed climate action in companies and industrial organizations through community building and support.[108]

Conclusions and Questions for Further Research

It is impossible to say definitively whether digital technology helps climate activism more than it hinders it. Technologies are not disembodied from wider social and political contexts; the balance of political and economic power and ambition is decisive for their consequences. What is clear is that a digitally connected world creates new opportunities for effective climate action as well as new risks arising from, inter alia, mass production and use of digital devices, large-scale surveillance, and new social divides between digital haves and have-nots. Thus, in incorporating digital technologies into their strategic repertoires there is a danger that climate activists inadvertently contribute to the problem they are fighting. Avoiding this danger requires a better understanding of how different digital technologies interact with different aspects of climate activism. Debates about the effects of social media must push beyond simple dichotomies between benefits of scale versus drawbacks of slacktivism and digital repression towards systematic theorising and empirical testing of specific mechanisms and claims (Lynch, 2011). This new research agenda should harness new methods to analyse the vast amounts of data available online to better understand the effects of online mobilization and government crackdowns. While there is a burgeoning literature on online repression in authoritarian regimes, there are fewer studies of digital repression in democratic regimes, or of how digital repression specifically affects climate activism. To what extent are climate activists in different parts of the world resilient to online censorship? When does censorship create backlash? Do climate activists face different challenges from other groups? Scholars also need to gain a better understanding of how different uses of technology affect the climate footprint of organizations to promote a focus on more sustainable ways of connecting. Finally, research focused on understanding when and why the backlash against climate action may prove stronger than the campaign for positive change will be crucial to effective resistance to online disinformation.

---

[108] https://climateaction.tech/community/.

Conclusion

Do digital technologies improve the impact of climate change activism? Some scholars and commentators argue that new digital technologies are causing fundamental societal change, wiping away old social and organizational structures and patterns of action. Yet, fundamental societal changes are rarely driven by technological innovation alone but arise from the way(s) in which new technologies interact with existing social, political, and economic structures to produce change. In this Element we have therefore focused on how specific technologies are influencing and changing specific aspects of climate activism; from mobilization and campaigning to monitoring and enforcement, organizational formation and fundraising, and elite lobbying.

Internet and social media have significantly lowered costs to climate activists of organizing, fundraising, mobilizing support, and networking domestically and internationally. New technologies have also led to improvements in elite lobbying, and to more effective monitoring of state compliance with national and international climate pledges. More broadly, digital technologies have contributed to challenging the monopoly of the state over critical information and data infrastructure (Beraldo and Milan, 2019). Climate activists today have a wealth of data at their command; from satellite, and other remote sensing data, to open-source data on climate litigation and environmental policies, which they can use to better understand the causes and impacts of climate change, and to campaign more effectively for pro-climate political action. It is important to acknowledge, however, that not all climate organizations are equally able to harness data analytics and satellite data; a point we return to later.

In addition to seeing new forms of digital activism, we have also seen a wealth of new climate activist organizations forming in the past decade, including many youth organizations such as Sunrise, and Fridays For Future. While none of these groups are defined by their use of digital technology or social media, they have benefitted greatly from it in terms of attracting members, fundraising, and coordinating collective action. In sum, technology has both *supersized* climate activism (enabling existing organizations to quickly scale up their campaigns, lobbying, and monitoring activities, and enabling new organizations to form) and has also *transformed* how some organizations campaign (for example, enabling organizations to hand over agenda-setting power to supporters, to crowdsource funding, or to enlist citizen scientists to scale up and democratize environmental research).

Who Uses Tech for What?

As we have illustrated in previous sections, climate organizations differ widely in the extent to which they rely on modern technology and for what purposes.

Such variation may be idiosyncratic, but there may also be underlying factors shaping technology use including, for example, political context, organizational size, age, and funding.

Generalist versus Specialist Organizations

Much scholarship on transnational advocacy has focused on large, well-known organizations whose work has global reach. These 'household-name' NGOs tend to have a 'generalist' profile, meaning they work across a broad range of issues and geographic areas, and tend to focus on mainstream issues that appeal to large public constituencies and are deemed acceptable by powerful policymakers (Stroup and Wong, 2017). However, recent studies of the ecologies of NGO populations have shown that global NGO populations are increasingly dominated by smaller, specialist organizations whose missions are narrower and who use a more limited set of strategies and tactics (Eilstrup-Sangiovanni 2019; Bush and Hadden, 2019). Although more research is needed to establish a clear pattern, these studies suggest that technologies tend to be used differently by generalist and specialist organizations. For generalist organizations, digital technology often presents a tool to supercharge existing strategic repertoires by increasing the scope and speed, and lowering the cost, of familiar practices. For smaller, specialist organizations, however, technology has often been found to serve as a basis for innovation of new strategies and tactical repertoires, such as technology-assisted monitoring and enforcement (Eilstrup-Sangiovanni, 2019; Eilstrup-Sangiovanni and Sharman, 2021; 2022).

Legacy NGOs versus Digital Natives Organizations

Similar to the divide between generalist and specialist organizations, scholars have found a digital divide between older (and larger) NGOs and CSOs and younger (smaller) newcomers. Large legacy NGOs tend to have more resources to invest in digital strategies but are more likely to opt for broadcasting strategies (i.e., posting their own content) rather than using digital tools in more transformative ways. In contrast, smaller and younger NGOs are found to use their online presence to engage more directly with their supporters and to network with other groups (Mitchell et al., 2020). As legacy NGOs such as WWF, Friends of the Earth, and Greenpeace have come under pressure from new digital advocacy organizations, many have responded by increasing their online presence (Hall et al., 2020; Hall et al., 2020). This is not surprising; after all, most NGOs use digital platforms to educate members and the public about their cause. However, NGOs founded before the internet era often struggle to embrace more transformative digital strategies such as 'conversing', 'digital analytics'/'testing', or 'facilitating' because it challenges their long-established staff-led, and expertise-driven organizational structures (Hall et al., 2019). There are some exceptions: Greenpeace, for example, created the Mobilization Lab to stimulate more member-driven campaigning within the organization, and many national Greenpeace sections have online petitions where supporters can initiate their own campaigns (Hall et al., 2020). Overall, however, legacy NGOs with professionalized and staff-led advocacy strategies are less likely than digitally native organizations to cede substantial control over campaigns to supporters, whereas digital natives are more open to online feedback and supporter-led actions, and more likely to embrace digital analytics (Hall et al., 2020).

Global North versus Global South

Given uneven access to digital infrastructures, and high demands on technical skills, one might assume that NGOs in the Global North, which tend to be relatively better funded, are more likely to adopt new digital technologies than those based in the Global South. However, NGOs in the Global South often 'leapfrog' their northern counterparts in digital innovation. For example, NGOs in India and South Africa have been faster to use SMS and peer-to-peer text messaging than many campaigning NGOs in the Global North who are more reliant on email for communication (Hall, 2022). When thinking about digital divides in the context of wider organizational ecologies, we must also consider how global demographic changes affect who is using digital technologies, and how. In 1996, in the early days of the internet, 80 percent of internet users were based in North America and Europe; today two-thirds of internet users are in the Global South (Kapur, 2024:58). As Kapur points out, 'India and China now account for about half the world's mobile data traffic; the fastest-growing population of users is in Africa'. The technologies that have worked for climate NGOs in the Global North may not be the same as those that will be embraced in the Global South, especially if we end up with different governance models for the internet in different parts of the world, as some predict (Bradford, 2023).

Varying Responses to Surveillance, Repression, and Backlash

Who uses tech for what will often depend on the political context in which climate organizations are operating. As discussed in earlier sections, states are increasingly using digital technology to monitor and repress activists (Gohdes, 2024). Research suggests that both democratic and authoritarian regimes are shrinking civic space and making it more difficult for activists to fundraise, network, and coordinate internationally (Chaudhry, 2022). Digital repression comes in many forms – from internet shutdowns, to targeted surveillance (Gohdes, 2024; Earl et al., 2022). Often activists find ways to navigate repression and backlash, either by opting for non-digital forms of communication and focusing their efforts on off-line activities, or by developing new technological solutions to avoid surveillance. These choices in turn depend on what forms of digital repression activists encounter, and what civic and political rights and avenues of juridical protection are available to them to seek redress.

We see great potential for further research into which types of climate advocacy organizations use which technologies, and for what purposes. Specifically, we see a need for research comparing technology use by older legacy NGOs with younger digital natives such as Fridays For Future, XR, 350.org, Last Generation, and Sunrise. Scholars should also continue to build on organizational ecology literature to understand patterns of specialization and 'niche-seeking' among climate advocacy groups (Eilstrup-Sangiovanni, 2019; Eilstrup-Sangiovanni and Sharman, 2022; Bush and Hadden, 2019), and to investigate which organizations are most likely to innovate, under what conditions. We also see a need for more systematic research to understand how digital repression influences tech use by climate groups in different geographic, political, and cultural settings. Overall, more comparative studies are needed to understand to what extent different climate organizations are specializing in using specific technologies to step up their fight against climate change.

Drivers of Strategic Innovation

In addition to understanding ecologies of NGO tech use, there is also a need to understand where the impetus for technological innovation comes from, and to scrutinize the growing relationship between the for-profit tech sector and the non-profit world of climate activism. As we have pointed out, many of the technological innovations that climate activist organizations use on a daily basis (whether satellite data, AI, or Facebook, or specific platforms and applications such as Climate Action Tracker or Global Forest Watch) are developed, funded, and serviced by the private corporate sector. Recent years have seen the emergence of communities of tech experts who specialize in developing software for non-profits and advocacy organizations, or who develop and manufacture purpose-built technological tools for NGOs such as 'eco-surveillance drones'. Although many 'tech-for-good' developers in the climate sector operate on a non-profit basis, they too must find ways to fund their initiatives. We do not have sufficient data to point to a clear trend, but anecdotal evidence from many of our examples and interviews suggests that many operate on start-up grants from Google.org and other big tech companies. If this is a firm trend, it raises thorny questions about whether the development of many new technologies supporting climate monitoring and advocacy are, at least indirectly, driven and shaped by incentives to stimulate technological innovation with potential for commercial application.

Of course, big tech companies may also sponsor technology for use by climate NGOs with less nefarious motives, such as showcasing the public benefits generated by their business. One (anonymous) interviewee suggested that Google's generous sponsorship of platforms like the Indigenous Mapping Platform, SkyTruth, and ClimateTRACE (among many others) is ultimately a form of CSR policy, or even an 'employee benefit' in that the opportunity to dedicate time and resources to develop technologies 'for good' makes tech developers employed at Google feel more positive about their jobs. If, at the same time, such activities spur creative innovation from which Google ultimately benefits commercially ... well, that's clearly a win-win, they suggested. We believe that scholarship on climate advocacy would benefit from looking closer into the nexus between the incentives and actions of the corporate tech sector and technological innovations in the climate advocacy sector. We should also pay close attention to whether and how big tech companies tackle their own climate emissions, and address problems of 'extractivism' in the industry.

Technological change is not the only explanation for strategic innovation among climate advocacy organizations. A sense that 'time is running out' is central to powering contemporary climate activism, perhaps more so than for other environmental problems due to the perceived urgency of the problem and irreversible 'tipping points' in the climate system. This prompts some activists to take a more radical approach, as explicitly urged by Malm (2021) in *How to Blow Up a Pipeline*. A sense of urgency may also play a role in driving activists to form new organizations as they grow frustrated with the tactics and/or lack of progress of existing environmental NGOs and movements. As the climate crisis has worsened, activists have also broadened their goals, from a primary focus on mitigation (reducing emissions), to include loss and damage, adaptation, and climate justice (Allan, 2021; Vanhala, forthcoming). Technology may be used in different ways by these broader climate movements.

In addition to technological enablers and a growing sense that time is running out, the goals and strategies of climate activist organizations are also shaped by economic shocks and political crises. The global financial crisis, the recession that followed it, and the subsequent COVID pandemic led governments to bail out certain sectors of the economy. In the United States and the EU, legislators proposed 'Green New Deals' to stimulate green growth (Moschella, 2024). Many governments are taking a more proactive role in incentivizing and financing climate action, compared with twenty years ago. This may in turn create more opportunities and openings for climate activists to push for state-financed climate action, whether through 'inside' elite lobbying or 'outside' mobilization of public pressure.[109] At the same time, we have also seen geopolitical tensions spark new conflicts, leading many governments to increase their military budgets. Militarization and conflict contribute significantly to climate change. As Crawford (2022) reports, the US military is the largest single institutional contributor of GHG emissions in the world. Conflicts may also distract state leaders from tackling climate issues and narrow opportunities for activists to push for political action. Although they go far beyond the scope of this Element, these wider structural changes in global politics influence climate advocacy in profound ways. It goes without saying that organizational innovation and change is not solely driven by technological changes.

Looking beyond climate activism, there is more research to be done on how technology is challenging and enabling national and transnational advocacy more broadly. While some of the technology uses we highlight in this Element may be specific to climate change activism, many of the tech-based challenges and opportunities we have identified also apply beyond the climate movement to activists working for human rights, women's and children's rights, disarmament, migrant and refugee rights, and so on. Like pro-climate organizations, many of these movements are facing strong opposition from far-right movements that are also harnessing digital technologies, in particular social media, to further their cause. How activists navigate this backlash both online and offline is an important question for scholars and practitioners alike.

Implications for Practitioners and Activists

We wrote this short Element in the hope that it would be of interest to both students and scholars of non-state activism broadly, and to activists themselves. At the time of writing, AI and ChatGPT are big new innovations; however, technological change can move fast. Hence it would be problematic for us to suggest specific technological solutions or strategies for climate activists. Rather we hope this Element will encourage practitioners to think critically about how technology can support or transform their work. Technology can both rationalize and scale up existing work – whether this be fundraising, mobilizing members, or collecting data and monitoring states – and offer activist organizations completely new ways of working, for example, by handing power to members to initiate online campaigns, and empowering citizen scientists to evidence environmental damage and hold their governments to account. In considering how to embrace technology, activist organizations should foreground their *theory of change* (i.e., their strategy for effecting

---

[109] We have also seen the rise of 'shareholder' activism with people buying stocks in companies, and then holding them to account on their climate emissions.

change in the world) and consider if and how different technologies may enable them to do this better. They should also be looking outwards and learning from other organizations that have different theories of change. There may be opportunities for collaboration between niche and generalist groups.

This Element has also raised wider questions about the implications of digital technologies for activists. Organizations that use technology to optimize and/or scale up their work may become more 'efficient' at producing campaign content, but no more effective at influencing decision-makers and public opinion. Hence, organizations should consider how and when to engage digital specialists in organizational decision-making. Digital experts can be consulted simply to maximize views, donations, or reach online; or they can be included in strategic decision-making to look for transformative ways to promote climate action. If activists have a deep understanding of their organizational mission and theory of change, as well as an understanding of the technological opportunities available, they may find novel ways to promote climate action. There are also specialist organizations that bridge the advocacy and tech worlds (such as Access Now and Tactical Tech) and provide important advice on how to navigate government repression and backlash. The support of such organizations can be critical especially for activist organizations that frequently cross international borders and/or enter new contexts – such as those attending recent UNFCCC Conferences of the Parties (COP) in Dubai or Azerbaijan.

The question of whether, and how, new technologies can help and/or hinder climate activism will remain important, as new technologies emerge, and global warming continues. We hope this Element will contribute to the ongoing conversation.

References

Aday, S., & Livingston, S. 2009. NGOs as Intelligence Agencies: The Empowerment of Transnational Advocacy Networks and the Media by Commercial Remote Sensing in the Case of Iran. *Geoforum*, **40**(3), 514–522.

Aday, S., & Livingston, S. 2016. *NGOs as Intelligence Agencies*. DOI: https://doi.org/10.1016/j.geoforum.2008.12.006.

Aguilar, D. 2018. Ecuador: Waorani People Map Their Rainforest to Save It. *Mongabay*, 6 June. https://news.mongabay.com/2018/06/ecuador-waorani-people-map-their-rainforest-to-save-it/.

Allan, J. 2021. *The New Climate Activism: NGO Authority and Participation in Climate Change Governance*. Toronto: University of Toronto Press.

Andrae, A., & Edler, T. 2015. On Global Electricity Usage of Communication Technology: Trends to 2030. *Challenges*, **6**(1), 117–157.

Baker, J. C., & Williamson, R. A. (2006). *Satellite Imagery Activism: Sharpening the Focus on Tropical Deforestation*. Santa Monica, CA: RAND Corporation.

Baran, Z., & Stoltenberg, D. 2023. Tracing the Emergent Field of Digital Environmental and Climate Activism Research: A Mixed-Methods Systematic Literature Review. *Environmental Communication*, **17**(5), 453–468.

Bennett, M. M., Chen, J. K., Alvarez León, L., & Gleason, C. J. 2022. The Politics of Pixels: A Review and Agenda for Critical Remote Sensing. *Progress in Human Geography*, **46**(3), 729–752. DOI: https://doi.org/10.1177/03091325221074691.

Bennett, W. L., & Segerberg, A. 2012. The Logic of Connective Action. *Information, Communication & Society*, **15**(5), 739–768.

Bennett, W. L., & Segerberg, A. 2013. *The Logic of Connective Action: Digital Media and the Personalization of Contentious Politics*. Cambridge, UK: Cambridge University Press.

Beraldo, D., & Milan, S. 2019. From Data Politics to the Contentious Politics of Data. *Big Data & Society*, **6**(2), 205395171988596. DOI: https://doi.org/10.1177/2053951719885967.

Berry, Jeffrey M. *Lobbying for the People: The Political Behavior of Public Interest Groups*. 1977. Princeton, NJ: Princeton University Press. http://www.jstor.org/stable/j.ctt13x11mt.

Bitonti, A. 2024. Tools of Digital Innovation in Public Affairs Management: A Practice-Oriented Analysis, *Journal of Public Affairs* **24**(1), e2888. DOI: https://doi.org/10.1002/pa.2888.

Bladen, Sarah. 2018. Close Encounters of the Fishy Kind. *Global Fishing Watch*, 8 June 2018. https://globalfishingwatch.org/news-views/close-encounters-of-the-fishy-kind/.

Bond, B., & Exley, Z. 2016. *Rules for Revolutionaries.* Chelsea, VT: Chelsea Green Publishing.

Bouwer, K. (2020). Lessons from a Distorted Metaphor: The Holy Grail of Climate Litigation. *Transnational Environmental Law*, 9(2), 347–378.

Bouwer, K., & Setzer, J. 2020. New Trends in International Climate and Environmental Advocacy. Paper presented at Workshop on New Trends in Climate Litigation: What Works? Bologna, May 15.

Bradford, A. 2023. *Digital Empires: The Global Battle to Regulate Technology.* Oxford: Oxford University Press.

Brulle, R. J. 2018. The Climate Lobby: A Sectoral Analysis of Lobbying Spending on Climate Change in the USA, 2000 to 2016. *Climatic Change*, 149(3–4), 289–303. https://link.springer.com/article/10.1007/s10584-018-2241-z

Brulle, R., & Downie, C. 2022. Following The Money: Trade Associations, Political Activity and Climate Change. *Climatic Change*, 175, 11. https://link.springer.com/article/10.1007/s10584-022-03466-0.

Bush, S., & Hadden, J. 2019. Density and Decline in Founding of INGOs in the United States. *International Studies Quarterly*, 63(4), 1133–1146.

Castells, M. 2012. *Networks of Outrage and Hope: Social Movements in the Internet Age.* Cambridge, UK: Polity Press.

Center for Countering Digital Hate. 2022. *Facebook Failing to Flag Harmful Climate Misinformation, New Research Finds.* https://counterhate.com.

Chalmers, A. W., & Shotton, P. A. (2016). Changing the Face of Advocacy? Explaining Interest Organizations' Use of Social Media Strategies. *Political Communication*, 33(3), 374–391. DOI: https://doi.org/10.1080/10584609.2015.1043477.

Chapin, M., Lamb, Z., & Threlkeld, B. 2005. Mapping Indigenous Lands. *The Annual Review of Anthropology*, 34(6), 19–38.

Chaudhry, S. 2022. The Assault on Civil Society: Explaining State Crackdown on NGOs. *International Organization*, 76(3), 549–590.

Chaudhry, S., & Heiss, A. 2022. Closing Space and the Restructuring of Global Activism: Causes and Consequences of the Global Crackdown on NGOs. Chapter 2 in *Beyond the Boomerang: From Transnational Advocacy Networks to Transcalar Advocacy in International Politics*, ed. Christopher L. Pallas and Elizabeth Bloodgood. Tuscaloosa, AL: University of Alabama Press.

Cheon, A., & Urpelainnen, J. 2018. *Activism and the Fossil Fuel Industry.* Abingdon: Routledge.

Cifuentes, S. 2020. Forest Monitoring Programmes and Indigenous Autonomy in the Amazon Basin. In *Global Information Society Watch Report: Technology, the Environment and a Sustainable World: Responses from the Global South*, ed. Alan Finlay, pp. 170–173. Association for Progressive Communications (APC), www.apc.org/en.

Climate Action Against Disinformation (CAAD). 2023. *Deny Delay Distort (Vol.3): Climate Information Integrity ahead of COP28*. https://caad.info/ analysis/reports/deny-deceive-delay-vol-3-climate-information-integrity-ahead-of-cop28/.

Climate Council. 2023. Driving Climate Action from Day One. *Climate Council*. 14 July. www.climatecouncil.org.au/our-impact-day-one/.

Cluverius, J. 2015. Grassroots Lobbying and the Economics of Political Information in the Digital Age. Unpublished dissertation, University of North Carolina, Chapel Hill.

Crawford, N. 2021. *Atlas of AI*. New Haven, CT: Yale University Press.

Crawford, N. 2022. *The Pentagon, Climate Change and War*. Cambridge, MA: Massachusetts Institute of Technology Press.

Cukier, K., & Mayer-Schoenberger, V. 2013. The Rise of Big Data: How It's Changing the Way We Think about the World. *Foreign Affairs*, **92**(3), 28–40.

DataReportal. 2024. *Digital 2024: Global Overview Report*. https://datarepor tal.com/reports/digital-2024-global-overview-report.

Dauvergne, P. 2020. *AI in the Wild: Sustainability in the Age of Artificial Intelligence*. Cambridge, MA: Massachusetts Institute of Technology Press.

Dawes, S. S., Vidiasova, L., & Parkhimovich, O. 2016. Planning and Designing Open Government Data Programs: An Ecosystem Approach. *Government Information Quarterly*. **33**(1), 15–27.

Dellmuth, L. M., & Bloodgood, E. A. 2019. Advocacy Group Effects in Global Governance: Populations, Strategies, and Political Opportunity Structures. *Interest Groups & Advocacy*, **8**, 255–269. DOI: https://doi.org/10.1057/ s41309-019-00068-7.

Dellmuth, L., & Tallberg, J. 2017. Advocacy Strategies in Global Governance: Inside versus Outside Lobbying. *Political Studies*, **65**(3), 705–723.

de Moor, J. 2023. Time and Place in Climate Activism: Three Urgency-Induced Debates. In Jeannie Sowers, Stacy D. VanDeveer, and Erika Weinthal (eds.), *The Oxford Handbook of Comparative Environmental Politics*, pp. 299–316 (online ed., Oxford Academic, 14 July 2021). DOI: https://doi.org/10.1093/ oxfordhb/9780197515037.013.31.

de Moor, J., De Vydt, M., Uba, K., & Wahlström, M. (2020). New Kids on the Block: Taking Stock of the Recent Cycle of Climate Activism. *Social*

Movement Studies, **20**(5), 619–625. DOI: https://doi.org/10.1080/14742837 .2020.1836617.

Dennis, J. 2019. *Beyond Slacktivism: Political Participation on Social Media.* Cham: Palgrave Macmillan.

Dietz, M., & Garrelts, H. 2014. *Routledge Handbook of the Climate Change Movement.* Abingdon: Routledge.

Dobson, A. 2016. *Environmental Politics: A Very Short Introduction.* Oxford: Oxford University Press.

Downie, C. 2023. Briefing Note: The Influence of the Fossil Fuel Sector. Workshop held at The Intellectual Forum, Jesus College Cambridge, June. https://www.jesus.cam.ac.uk/sites/default/files/Briefing%20from%20FF% 20workshop%20Jesus%20College%20Cambridge.pdf.

Drieschova, A. 2023. The Social Media Revolution and Shifts in the Climate Change Discourse. Chapter 9 in *Digital International Relations Technology, Agency and Order*, ed. Corneliu Bjola and Markus Kornprobst. London: Routledge.

Earl, J., Maher, T. V., & Pan, J. 2022. The Digital Repression of Social Movements, Protest, and Activism: A Synthetic Review. *Science Advances*, **8**(10). DOI: https://doi.org/10.1126/sciadv.abl8198.

Eilstrup-Sangiovanni, M. 2019. Competition and Strategic Differentiation among Transnational Advocacy Groups. *Journal of Interest Groups and Advocacy*, **8**(3), 376–406.

Eilstrup-Sangiovanni, M., & Sharman, J. C. 2021. Enforcers beyond Borders: Transnational NGOs and the Enforcement of International Law. *Perspectives on Politics.* **19**(1), 131–147. DOI: https://doi.org/10.1017/S153759271900344X.

Eilstrup-Sangiovanni, M. & Sharman, J. C. 2022. *Vigilantes beyond Borders.* Princeton, NJ: Princeton University Press.

Engine Room. 2023. Using Geographic Information Systems (GIS) to Gather, Manage, Analyse and Visualize Spatial and Geographic Data. www.theengi neroom.org/understanding-the-impact-of-geospatial-data-in-social-and-cli mate-justice.

Farrell, H. 2012. The Consequences of the Internet for Politics. *Annual Review of Political Science*, **15**(1), 35–52.

Feldstein, S. 2021. *The Rise of Digital Repression: How Technology Is Reshaping Power, Politics, and Resistance.* Oxford: Oxford University Press.

Fisher, D. R. 2019. *American Resistance: From the Women's March to the Blue Wave.* New York, NY: Columbia University Press.

Fisher, D. R., & Nasrin, S. 2021. Climate Activism and Its Effects. *WIREs Climate Change*, **12**(1), e683. DOI: https://doi.org/10.1002/wcc.683.

Fung, A., & O'Rourke, D. 2000. Reinventing Environmental Regulation from the Grassroots Up: Explaining and Expanding the Success of the Toxics Release Inventory. *Environmental Management*, **25**, 115–127. https://doi.org/10.1007/s002679910009

Fung, A., Russon Gilman, H., & Shkabatur, J. 2013. Six Models for the Internet + Politics. *International Studies Review*, **15**(1), 30–47.

Fung, E. 2022. How Digitized Strategy Impacts Movement Outcomes: Social Media, Mobilizing, and Organizing in the 2018 Teachers' Strikes. *Politics & Society*, **50**(3), 458–518.

Funk, A., Shahbaz, A., & Vesteinsson, K. 2023. Freedom on the Net 2023: The Repressive Power of Artificial Intelligence. *Freedom House*. https://freedomhouse.org/report/freedom-net/2023/repressive-power-artificial-intelligence.

Gerbaudo, P. 2012. *Tweets on the Streets: Social Media and Contemporary Activism*. London: Pluto Press.

Gerbaudo, P. 2018. *The Digital Party: Political Organisation and Online Democracy*. London: Pluto Press.

Gibson, R., Römmele, A., & Williamson, A. (2014). Chasing the Digital Wave: International Perspectives on the Growth of Online Campaigning. *Journal of Information Technology & Politics*, **11**(2), 123–129. DOI: https://doi.org/10.1080/19331681.2014.903064.

Gladwell, M. 2010. Small Change: Why the Revolution Will Not Be Tweeted. *The New Yorker*. www.newyorker.com/magazine/2010/10/04/small-change-malcolm-gladwell.

Glicksman, Robert L. and Markell, David L. and Monteleoni, Claire, *Technological Innovation, Data Analytics, and Environmental Enforcement (September 19, 2016). Ecology Law Quarterly, Vol. 44, No. 1; FSU College of Law, Public Law Research Paper No. 815; GWU Law School Public Law Research Paper No. 2016–48; GWU Legal Studies Research Paper No. 2016–48.* Available at SSRN: http://ssrn.com/abstract=2840944.

Glicksman, R. L., Markell, D. L., & Monteleoni, C. (2016). Technological Innovation, Data Analytics, and Environmental Enforcement. *Ecology Law Quarterly*, **44**(1), 12.

Global Information Society Watch. 2020. *Technology, the Environment, and a Sustainable World: Responses from the Global South*. https://giswatch.org/2020-technology-environment-and-sustainable-world-responses-global-south.

Global Witness. 2020. *Beef, Banks and the Brazilian Amazon*. www.globalwitness.org/en/campaigns/forests/beef-banks-and-brazilian-amazon/.

Gohdes, A. 2020. Repression Technology: Internet Accessibility and State Violence. *American Journal of Political Science*, **64**(2), 488–503.

Gohdes, A. 2024. *Repression in the Digital Age: Surveillance, Censorship, and the Dynamics of State Violence*. Oxford: Oxford University Press.

Grygiel, J. 2019. Facebook Algorithm Changes Suppressed Journalism and Meddled with Democracy. *The Conversation*. http://theconversation.com/facebook-algorithm-changes-suppressed-journalism-and-meddled-with-democracy-119446.

Gunitsky, S. 2015. Corrupting the Cyber-Commons: Social Media as a Tool of Autocratic Stability. *Perspectives on Politics*, **13**(1), 42–54.

Gutiérrez, M. 2018. *Data Activism and Social Change*. Palgrave Studies in Communication for Social Change, ed. P. Thomas and E. van der Fliert. Heidelberg: Springer. https://link.springer.com/book/10.1007/978-3-319-78319-2.

Gutiérrez, M. 2019. Maputopias: Cartographies of Knowledge, Communication and Action in the Big Data Society – the Cases of Ushahidi and InfoAmazonia. *GeoJournal*, **84**, 101–120. DOI: https://doi.org/10.1007/s10708-018-9853-8.

Hadden, J., & Jasny, L. 2017. The Power of Peers: How Transnational Advocacy Networks Shape NGO Strategies on Climate Change. *British Journal of Political Science*, **49**(2), 637–659. DOI: https://doi.org/10.1017/S0007123416000582.

Hall, N. 2022. *Transnational Advocacy in the Digital Era, Think Global, Act Local*. Oxford: Oxford University Press.

Hall, N., Schmitz, H. P., & Dedmon, J. M. 2020. Transnational Advocacy and NGOs in the Digital Era: New Forms of Networked Power. *International Studies Quarterly*, **64**(1), 159–167.

Han, H. 2014. *How Organizations Develop Activists: Civic Associations and Leadership in the 21st Century*. Oxford: Oxford University Press.

Hestres, L. E. 2015. Climate Change Advocacy Online: Theories of Change, Target Audiences, and Online Strategy. *Environmental Politics*, **24**(2), 193–211. DOI: https://doi.org/10.1080/09644016.2015.992600.

Hestres, Luis E., & Jill E. Hopke. 2016. *Internet-Enabled Activism and Climate Chance*. Oxford: Oxford University Press.

Higgins, E. 2021. *We Are Bellingcat*. London: Bloomsbury.

Hobbs, W. R. & Roberts, M. E. 2018. How Sudden Censorship Can Increase Access to Information. *American Political Science Review*, **112**(3), 621–636.

Holtz, C., Silberman, M., & Mahendra, J. 2015. Beyond Vanity Metrics: Toward Better Measurement of Member Engagement. *MobLab*. https://mobilisationlab.org/resources/beyond-vanity-metrics-toward-better-measurement-of-member-engagement/.

Hong Tien Vu, Hung Viet Do, Hyunjin Seo & Yuchen Liu 2020. Who Leads the Conversation on Climate Change?: A Study of a Global Network of NGOs on Twitter. *Environmental Communication*, **14**(4), 450–464.

Howe, J. 2006. The Rise of Crowdsourcing. *Wired Magazine*, **14**, 1–4.

Humby, T-L. 2018. The Thabametsi Case: Case No 65662/16 Earthlife Africa Johannesburg v Minister of Environmental Affairs. *Journal of Environmental Law*, **30**(1), 145–155. https://doi.org/10.1093/jel/eqy007

Ijjasz-Vasquez, E., & Coughenour Betancourt, A. 2023. How Technology Can Help Save Indigenous Forests – and Our Planet. *Eco-business*. www.eco-business.com/opinion/how-technology-can-help-save-indigenous-forests-and-our-planet/.

InfluenceMap. 2024. https://influencemap.org.

IPCC. 2022. *Climate Change 2022: Impacts, Adaptation and Vulnerability*. IPCC Sixth Assessment Report. https://www.ipcc.ch/report/ar6/wg2/.

Isaac, M., & Ember, S. 2016. Facebook to Change News Feed to Focus on Friends and Family. *The New York Times*. www.nytimes.com/2016/06/30/technology/facebook-to-change-news-feed-to-focus-on-friends-and-family.html.

Kapur, A. 2024. Prometheus Bound. *The New Yorker*, 5 February, 53–58.

Karpf, D. 2012. *The MoveOn Effect: The Unexpected Transformation of American Political Advocacy*. New York: Oxford University Press.

Karpf, D. 2016. *Analytic Activism*. Oxford: Oxford University Press.

Karpf, D. 2020. Two Provocations for the Study of Digital Politics in Time. *Journal of Information Technology & Politics*, **17**(2), 87–96.

Kazansky, B., Karak, N., Perosa, T., Tsuri, Q., Baker, S., & the Engine Room. 2022. At the Confluence of Digital Rights and Climate & Environmental Justice: A Landscape Review. https://engn.it/climatejusticedigitalrights

Keck, M. E., & Sikkink, K. 1998. *Activists beyond Borders: Advocacy Networks in International Politics*. Ithaca, NY: Cornell University Press.

Kingston, L. N., & Stam, K. R. 2013. Online Advocacy: Analysis of Human Rights NGO Websites. *Journal of Human Rights Practice*, **5**(1), 75–95.

Kirilenko, A., Desell, T., Kim, H., & Stepchenkova, S. 2017. Crowdsourcing Analysis of Twitter Data on Climate Change: Paid Workers vs. Volunteers. *Sustainability*, **9**, 1–15.

Koenig, Alexa. 2017. *Harnessing Social Media as Evidence of Grave International Crimes*. https://medium.com/humanrightscenter/harnessing-social-media-as-evidence-of-grave-international-crimes-d7f3e86240d.

Kreiss, D. 2016. *Prototype Politics: Technology Intensive Campaigning and the Data of Democracy*. New York: Oxford University Press.

Langer, M., & Eason, M. 2019. The Quiet Expansion of Universal Jurisdiction. *European Journal of International Law*, **30**(3), 779–817.

Li, P., Yang, J., Islam, M. A., & Ren, S. 2023. Making AI Less 'Thirsty': Uncovering and Addressing the Secret Water Footprint of AI Models. arXiv:2304.03271v3 [cs.LG]

Liu, Y., Li, W., and Wang, L. 2023. Why Greenwashing Occurs and What Happens Afterwards? A Systematic Literature Review and Future Research Agenda. *Environmental Science and Pollution Research*, **30**, 118102–118116.

Lowles, N. (Ed.). 2021. State of Hate 2021: Backlash, Conspiracies and Confrontation. *Hope not Hate*. https://hopenothate.org.uk/wp-content/uploads/2021/03/state-of-hate-2021-final-2.pdf

Lynch, M. 2011. After Egypt: The Limits and Promise of Online Challenges to the Authoritarian Arab State. *Perspectives on Politics*, **9**, 301–310. DOI: https://doi.org/10.1017/S1537592711000910. https://www.cambridge.org/core/journals/perspectives-on-politics/article/abs/after-egypt-the-limits-and-promise-of-online-challenges-to-the-authoritarian-arab-state/5CBBD764FBF392645FF7D53C1B0016FC.

Lythreatis, S., Singh, S. K., & El-Kassar, A.-N. 2022. The Digital Divide: A Review and Future Research Agenda. *Technological Forecasting and Social Change*, 175. DOI: https://doi.org/10.1016/j.techfore.2021.121359.

McLean, J. E., and Fuller, S. 2016. Action with(out) Activism: Understanding Digital Climate Change Action. *International Journal of Sociology and Social Policy*, **36**(9/10), 578–595.

Malm, A. 2021. *How to Blow Up a Pipeline*. London: Verso Books.

Margetts, H., John, P., Hale, S., & Yasseri, T. 2015. *Political Turbulence: How Social Media Shape Collective Action*. Princeton, NJ: Princeton University Press.

Merkel, W. 2017. The Limits of Democratic Innovations in Established Democracies. In *The Governance Report 2017: Democratic Innovations*, pp. 111–126. Oxford: Oxford University Press. www.hertie-school.org/en/governancereport/govreport-2017.

Milan, S. 2018. Data Activism as the New Frontier of Media Activism. In *Media Activism in the Digital Age: Charting an Evolving Field of Research*. London: Routledge.

Milan, S., & Velden, L. van der. The Alternative Epistemologies of Data Activism. In *Digital Culture & Society*, vol. 2, no. 2, 2016, pp. 57–74. https://doi.org/10.14361/dcs-2016-0205.

Mitchell, G. E., Schmitz, H. P., & Vijfeijken, T. B. 2020. *Between Power and Irrelevance: The Future of Transnational NGOs*. Oxford: Oxford University Press.

Mogus, J., Silberman, M., & Roy, C. 2011. Four Models for Managing Digital at Your Organization. *Stanford Social Innovation Review (SSIR)*. DOI: https://doi.org/10.48558/ZE2N-XC29.

Mogus, J., & Liacas, T. 2016. Networked Change: How Progressive Campaigns Are Won in the 21st Century. *NetChange Consulting Report*. https://commonslibrary.org/networked-change/.

Morozov, E. 2013. To Save Everything, Click Here: The Folly of Technological Solutionism.*Journal of Design History*, **27**(1), 111–113. DOI: https://doi.org/10.1093/jdh/ept034

Moschella, M. 2024. *Unexpected Revolutionaries: How Central Banks Made and Unmade Economic Orthodoxy*. Ithaca, NY: Cornell University Press.

Muller, C. L., Chapman, L., Johnston, S., Kidd, C., Illingworth, S., Foody, G., Overeem, A., & Graves, R. 2015. Crowdsourcing for Climate and Atmospheric Sciences: Current Status and Future Potential. *International Journal of Climatology*, **35**(11), 3185–3203. DOI: https://doi.org/10.1002/joc.4210.

Nakabuye, H. F., Nirere, S., & Oladosu, A. T. 2020. The Fridays For Future Movement in Uganda and Nigeria. In *Standing Up for a Sustainable World*. Cheltenham: Edward Elgar Publishing.

Neumayer, C., & Svensson, J. 2016, Activism and Radical Politics in the Digital Age: Towards a Typology. *Convergence: The International Journal of Research into New Media Technologies*, **22**(2), 131–146. DOI: https://doi.org/10.1177/1354856514553395.

Nonprofit Tech for Good. 2018. *Global Trends in Giving Report*. https://philanthropycircuit.org/reports/2018-global-trends-in-giving-report/.

Nonprofit Tech for Good. 2020. *Global Trends in Giving Report*. https://www.nptechforgood.com/2020-global-trends-in-giving-report/.

Nyabola, N. 2018. *Digital Democracy, Analogue Politics: How the Internet Era Is Transforming Kenya*. London: Zed Books.

Obar, J. A., Zube, P., & Lampe, C. (2012). Advocacy 2.0: An Analysis of How Advocacy Groups in the United States Perceive and Use Social Media as Tools for Facilitating Civic Engagement and Collective Action. *Journal of Information Policy*, **2**, 1–25. DOI: https://doi.org/10.5325/jinfopoli.2.2012.0001.

Ozden, J. 2022. Protest Movements Could Be More Effective Than the Best Charities. https://ssir.org/articles/entry/protest_movements_could_be_more_effective_than_the_best_charities.

Pallas, C. L., and Bloodgood, E. A. (eds.) 2022. *Beyond the Boomerang: From Transnational Advocacy Networks to Transcalar Advocacy in International Politics*. Tuscaloosa: University of Alabama Press.

Pan, J. & Siegel, A. A. 2020. How Saudi Crackdowns Fail to Silence Online Dissent. *American Political Science Review*, **114**, 109–125.

Paneque-Gálves, J., Vargas-Ramírez, N., Napoletano, B. M., & Cummings, A. 2017. Grassroots Innovation Using Drones for Indigenous Mapping and Monitoring. *Land*, **6**(4), 86. DOI: https://doi.org/10.3390/land6040086.

Peel, J., & Osofsky, H. M. 2018. Rights Turn in Climate Change Litigation? *Transnational Environmental Law*, **7**(1), 37–67.

Rahman, K. S. & Thelen, K. 2019. The Rise of the Platform Business Model and the Transformation of Twenty-First-Century Capitalism. *Politics & Society*, **47**(2), 177–204.

Rogers, R. 2018. Otherwise Engaged: Social Media from Vanity Metrics to Critical Analytics. *International Journal of Communication*, **12**, 459–460.

Rothe, D., & Shim, D. 2018. Sensing the Ground: On the Global Politics of Satellite-Based Activism. *Review of International Studies*, **44**(3), 414–437.

Saiger, A.-J. 2020. Domestic Courts and the Paris Agreement's Climate Goals: The Need for a Comparative Approach. *Transnational Environmental Law*, **9** (1), 37–54

Schradie, J. 2015. Labor Unions, Social Media, and Political Ideology: Using the Internet to Reach the Powerful or Mobilize the Powerless? *International Journal of Communication*, **9**, 21.

Schradie, J. 2018. Moral Monday Is More Than a Hashtag: The Strong Ties of Social Movement Emergence in the Digital Era. *Social Media + Society*, **4**(1), 205630511775071. DOI: https://doi.org/10.1177/2056305117750719.

Schradie, J. 2019. *The Revolution That Wasn't*. Cambridge: Harvard University Press.

Schulz, M. 2018. Why Those Holding the Purse Strings Must Do Data Better. *SmartyGrants*. https://smartygrants.com.au/articles/why-those-holding-the-purse-strings-must-do-data-better.

Scott, J. C. 2020. *Seeing Like a State*. New Haven, CT: Yale University Press. https://yalebooks.co.uk/book/9780300246759/seeing-like-a-state/

Setzer, J., & Higham, C. 2023. Global Trends in Climate Change Litigation: 2023 Snapshot. London: Grantham Research Institute on Climate Change and the Environment and Centre for Climate Change Economics and Policy, LSE.

Shirky, C. 2008. *Here Comes Everybody: The Power of Organizing Without Organizations*. London: Allen Lane.

Silberman, M., & Mahendra, J. 2015. Moving beyond Vanity Metrics. *Stanford Social Innovation Review (SSIR)*.

Sorce, G. 2023. Stuck with the Algorithm: Algorithmic Consciousness and Repertoire in Fridays For Future's Data Contention. *Media and Communication*, **11**(3), 214–225.

Stroup, S. S., & Wong, W. H. 2017. *The Authority Trap: Strategic Choices of International NGOs*. Ithaca, NY: Cornell University Press.

Stürmer, K., Rademacher, L., Fenton, P., & Suilleabhain, G. 2023. Discovering Digital Lobbying: How Digital Transformation and Social Media Affect Classic Lobbying Actors. In Rodríguez-Salcedo, N., Moreno, Á., Einwiller, S., & Recalde, M. (Ed.), *(Re)discovering the Human Element in Public Relations and Communication Management in Unpredictable Times*, **6**, 211–227.

Suman, B. 2018. Challenging Risk Governance Patterns through Citizen Sensing: The Schiphol Airport Case. *International Review of Law Computers & Technology*, **32**, 155–173.

Suman, B., & Schade, S. 2021. The Formosa Case: A Step Forward on the Acceptance of Citizen-Collected Evidence in Environmental Litigation? *Citizen Science: Theory and Practice*, **6**(1), 16.

Tattersall, A., Hinchliffe, J., & Yajman, V. 2022. School Strike for Climate Are Leading the Way: How Their People Power Strategies Are Generating Distinctive Pathways for Leadership Development. *Australian Journal of Environmental Education*, **38**(1), 40–56.

Thrall, A. T., Stecula, D., & Sweet, D. (2014). May We Have Your Attention Please? Human-Rights NGOs and the Problem of Global Communication. *The International Journal of Press/Politics*, **19**(2), 135–159.

Tilly, Charles. 1979. *From Mobilisation to Revolution*. Reading: Addison-Wesley.

Tufekci, Z. 2017. *Twitter and Tear Gas*. Oxford: Oxford University Press.

United Nations' High-Level Expert Group on the Net Zero Emissions Commitments of Non-State Entities. 2022. *Integrity Matters: Net Zero Commitments by Businesses, Financial Institutions, Cities and Regions*. https://www.un.org/sites/un2.un.org/files/high-level_expert_group_n7b.pdf.

Vanhala, L. Forthcoming. Governing the End: The Making of Climate Change Loss and Damage. Unpublished manuscript.

Vromen, A., Halpin, D., & Vaughan, M. 2022. *Crowdsourced Politics*. Basingstoke: Palgrave.

Walsh, J. 2024. *Digitally Networked Sleuthing: Online Platforms, Netizen Detectives, and Bottom-Up Investigators*. London: Routledge.

Wickens, M. P., & Louis, R. P. 2008. Mapping Indigenous Depth of Place. *American Indian Culture and Research Journal*, **32**(3), 107–126.

Willis, C. G., Law, E., Williams, A. C., et al. 2017. Crowdcurio: An Online Crowdsourcing Platform to Facilitate Climate Change Studies Using Herbarium Specimens. *New Phytologist*, **215**, 479–488.

Wong, W. 2012. *Internal Affairs: How the Structure of NGOs Transforms Human Rights*. Ithaca, NY: Cornell University Press.

World Bank. 2008. *The Role of Indigenous Peoples in Biodiversity Conservation*. https://documents1.worldbank.org/curated/en/995271468177530126/pdf/ 443000WP0BOX321onservation01PUBLIC1.pdf.

World Bank. 2023. *Digital Progress and Trend Report*. 2023. https://openknow ledge.worldbank.org/server/api/core/bitstreams/95fe55e9-f110-4ba8-933f-e65572e05395/content.

Wotzka, L. 2022. Can Mandatory Disclosure Help to Reduce Industrial Pollution? Assessing Stakeholder Use of the European Pollutant Release and Transfer Register in Germany. Master's thesis, University of Lund. https://lup.lub.lu.se/ luur/download?func=downloadFile&recordOId=9086537&fileOId=9086538.

Xu, J. & Zhang, H. 2022. Activating beyond Informing: Action-Oriented Utilization of WeChat by Chinese Environmental NGOs. *International Journal of Environmental Research and Public Health*, **19**(7), 3776.

Acknowledgements

We are grateful to Jennifer Hadden for inviting us to write this Element, to the CUP Elements editorial team for their assistance in publishing it, and to three anonymous reviewers for their constructive feedback. We also thank the many friends and colleagues who contributed along the way, including participants at the *Transnational Advocacy* conference in Bologna, May 2023; the *Environmental Politics and Governance* conference in Glasgow, July 2023; the Norwegian Center for Human Rights Conference on *Planetary Crisis and Human Rights* in Oslo, July 2023. Special thanks to Jean-Frédéric Morin, Aseem Prakash, Erika Weinthal, May Farid, and Anita Gohdes for their helpful comments on early drafts, and to Daniel Lubin and Josefine Petrick at SAIS Bologna and Zi Xuan Tang at Cambridge University for their excellent research assistance.

Cambridge Elements ⁼

Organizational Response to Climate Change

Aseem Prakash

University of Washington

Aseem Prakash is Professor of Political Science, the Walker Family Professor for the College of Arts and Sciences, and the Founding Director of the Center for Environmental Politics at University of Washington, Seattle. His recent awards include the American Political Science Association's 2020 Elinor Ostrom Career Achievement Award in recognition of "lifetime contribution to the study of science, technology, and environmental politics," the International Studies Association's 2019 Distinguished International Political Economy Scholar Award that recognizes "outstanding senior scholars whose influence and path-breaking intellectual work will continue to impact the field for years to come," and the European Consortium for Political Research Standing Group on Regulatory Governance's 2018 Regulatory Studies Development Award that recognizes a senior scholar who has made notable "contributions to the field of regulatory governance."

Jennifer Hadden

University of Maryland

Jennifer Hadden is Associate Professor in the Department of Government and Politics at the University of Maryland. She conducts research in international relations, environmental politics, network analysis, nonstate actors, and social movements. Her research has been published in various journals, including the *British Journal of Political Science, International Studies Quarterly, Global Environmental Politics, Environmental Politics,* and *Mobilization.* Dr. Hadden's award-winning book, *Networks in Contention: The Divisive Politics of Global Climate Change,* was published by Cambridge University Press in 2015. Her research has been supported by a Fulbright Fellowship, as well as grants from the National Science Foundation, the National Socio-Environmental Synthesis Center, and others. She held an International Affairs Fellowship from the Council on Foreign Relations for the 2015–16 academic year, supporting work on the Paris Climate Conference in the Office of the Special Envoy for Climate Change at the US Department of State.

David Konisky

Indiana University

David Konisky is Professor at the Paul H. O'Neill School of Public and Environmental Affairs, Indiana University, Bloomington. His research focuses on US environmental and energy policy, with particular emphasis on regulation, federalism and state politics, public opinion, and environmental justice. His research has been published in various journals, including the *American Journal of Political Science, Climatic Change,* the *Journal of Politics, Nature Energy,* and *Public Opinion Quarterly.* He has authored or edited six books on environmental politics and policy, including *Fifty Years at the U.S. Environmental Protection Agency: Progress, Retrenchment and Opportunities* (Rowman & Littlefield, 2020, with Jim Barnes and John D. Graham), *Failed Promises: Evaluating the Federal Government's Response to Environmental Justice* (MIT, 2015), and *Cheap and Clean: How Americans Think about Energy in the Age of Global Warming* (MIT, 2014, with Steve Ansolabehere). Konisky's research has been funded by the National Science Foundation, the Russell Sage Foundation, and the Alfred P. Sloan Foundation. Konisky is currently coeditor of *Environmental Politics.*

Matthew Potoski

UC Santa Barbara

Matthew Potoski is a Professor at UCSB's Bren School of Environmental Science and Management. He currently teaches courses on corporate environmental management, and his research focuses on management, voluntary environmental programs, and public policy. His research has appeared in business journals such as *Strategic Management Journal*, *Business Strategy and the Environment*, and the *Journal of Cleaner Production*, as well as public policy and management journals such as *Public Administration Review* and the *Journal of Policy Analysis and Management*. He coauthored *The Voluntary Environmentalists* (Cambridge, 2006) and *Complex Contracting* (Cambridge, 2014; the winner of the 2014 Best Book Award, American Society for Public Administration, Section on Public Administration Research) and was coeditor of *Voluntary Programs* (MIT, 2009). Professor Potoski is currently coeditor of the *Journal of Policy Analysis and Management* and the *International Public Management Journal*.

About the Series

How are governments, businesses, and nonprofits responding to the climate challenge in terms of what they do, how they function, and how they govern themselves? This series seeks to understand why and how they make these choices and with what consequence for the organization and the eco-system within which it functions.

Cambridge Elements ☰

Organizational Response to Climate Change

Elements in the Series

Explaining Transformative Change in ASEAN and EU Climate Policy: Multilevel Problems, Policies and Politics
Charanpal Bal, David Coen, Julia Kreienkamp, Paramitaningrum, and Tom Pegram

Fighting Climate Change through Shaming
Sharon Yadin

Governing Sea Level Rise in a Polycentric System: Easier Said than Done
Francesca Vantaggiato and Mark Lubell

Inside the IPCC: How Assessment Practices Shape Climate Knowledge
Jessica O'Reilly, Mark Vardy, Kari De Pryck, and Marcela da S. Feital Benedetti

Climate Activism, Digital Technologies, and Organizational Change
Mette Eilstrup-Sangiovanni and Nina Hall

A full series listing is available at: www.cambridge.org/ORCC

Printed in the United States
by Baker & Taylor Publisher Services